AF324890

India's Nuclear Energy Programme

*Published by*
ACADEMIC FOUNDATION
in association with

Indian National Science Academy

# India's Nuclear Energy Programme

## Future Plans, Prospects and Concerns

*Editor*

**R. Rajaraman**

ACADEMIC FOUNDATION

NEW DELHI

www.academicfoundation.com

First published in 2013
by

**ACADEMIC FOUNDATION**

4772-73 / 23  Bharat Ram Road, (23 Ansari Road),
Darya Ganj, New Delhi - 110 002 (India).
Phones : 23245001 / 02 / 03 / 04.
Fax : +91-11-23245005.
E-mail : books@academicfoundation.com
www.academicfoundation.com

*In association with*

Indian National Science Academy, New Delhi

*Disclaimer*:
The findings/views/opinions expressed in this book are solely
those of the authors and do not necessarily reflect the views of the
publisher.



Cataloging in Publication Data--DK
  Courtesy: D.K. Agencies (P) Ltd. <docinfo@dkagencies.com>

Workshop on Challenges in Nuclear Safety (2012 : Indian
  National Science Academy)
    India's nuclear energy programme : future plans, prospects
    and concerns / editor, R. Rajaraman.
      p.  cm.
      ISBN 9789332700307

      1. Nuclear energy--India--Safety measures--Congresses. 2.
  Nuclear energy--Technological innovations--India--
  Congresses. 3. Nuclear energy--Government policy--India--
  Congresses.   I. Rajaraman, R. (Ramamurti), 1939- II. Indian
  National Science Academy. III. Title.

DDC 333.79240954    23

Typeset by Italics India, New Delhi.

Printed and bound by The Book Mint, New Delhi.
www.thebookmint.in

# CONTENTS



## I

## Overview

1. The Nuclear Energy Option and
Nuclear Safety



2. Nuclear Power: Some Frequently Asked Questions



## II

## Reactor Safety and Radiological Concerns

3. Nuclear Reactor Safety



4. Environmental Impact on Aquatic Ecosystems,
Biodiversity, Agriculture and Human Health
in the Vicinity of Operating Nuclear Power Plants

## III

### Fuel Cycles and Technology



## IV

### Legal and Regulatory Issues

. . . . . . . . . . . . . . . . . . . . . . . . . . . . . . . . . . . . . . . . . . . . . . . . . . . . . . . . . . . . . . .

# List of Tables and Figures

### Tables

## Figures

# Editor/Contributors

*Editor:*

**R. Rajaraman** is an emeritus professor of theoretical physics at Jawaharlal Nehru University, New Delhi. He is Co-Chair of the International Panel on Fissile Materials and at the time of writing, Vice-President of the Indian National Science Academy. He got his PhD in 1963 from Cornell University under the supervision of the Nobel Laureate Prof Hans Bethe. Since then, he has been doing research and teaching in physics at Cornell University, the Institute for Advanced Study at Princeton, Harvard University, M.I.T, Stanford University, CERN, Delhi University and the Indian Institute for Science. His research areas include: nuclear theory, quantum field theory, statistical mechanics, particle physics and quantum hall systems. He has also been working on nuclear policy—both military and civilian. He has analysed India's nuclear doctrine, credible minimal deterrence, early warning systems, missile defence, battlefield nuclear weapons. With colleagues, he has extensively analysed fissile material production in India and Pakistan. He studied in depth the Indo-US nuclear agreement and its implications for India's civilian and strategic programs. Since Fukushima, he has also been addressing public concerns of reactor safety.

*Contributors:*

**Shree Kumar Apte** is a Distinguished Scientist and Associate Director (B), Biomedical Group, Bhabha Atomic Research Centre (BARC) at Mumbai. He obtained his Masters Degree from the Jiwaji University, Gwalior in 1973 and his PhD in Botany from Gujarat University. He joined BARC in 1974. For the last three decades, he has been supervising R&D programmes in stress biology and biotechnology at BARC. He was appointed Head, Molecular Biology Division in 2002. His work encompasses physiology, biochemistry, developmental biology, molecular genetics and biotechnology and addresses tolerance of bacteria and plants to major environmental stresses and their possible biotechnological applications in agriculture and environmental clean-up. His laboratory has successfully constructed India' first transgenic cyanobacteria capable of nitrogen fixation in stressful environments; and has engineered natural and recombinant strains of bacteria for removal of uranium from nuclear waste and seawater. Prof Apte co-ordinated, a 5-year duration, DAE sponsored multi-institutional project entitled: Thermal Ecological Studies which assessed the ecological impact of thermal effluents from operating nuclear power plants on aquatic ecosystems. He is a fellow of the INSA, National Academy of Sciences, National Academy of Agricultural Sciences, and Maharashtra Academy of Sciences. He was President of the Indian Society of Cell Biology during 2008-2010. He is recipient of Prof J.V. Bhat-Eureka Forbes Award for Excellence in Microbiology, 1990; Young Scientist Award in Cyanobacterial Biotechnology, 1996; Prof K.S. Bilgrami Memorial Award of INSA in 2006; INS-2006 award from the Indian Nuclear Society. He has been awarded international IAEA fellowship in 1976-77; the Nuffield Foundation Fellowship, U.K in 1984, the USAID Fellowship, in 1988-89. He is a DST J.C. Bose National Fellow. He is also a Senior Professor and Dean Academic, Life Sciences of the Homi Bhabha National Institute.

**Srikumar Banerjee** joined Bhabha Atomic Research Centre (BARC), Mumbai in 1967 after obtaining BTech in metallurgical engineering as well as his PhD from IIT, Kharagpur. As Director, BARC during 2004-2010, he provided leadership in the development of advanced reactor technology, fuel cycle technology and strategic programmes. He went on to become the Chairman, Atomic Energy Commission and Secretary to the Government of India, Department of Atomic Energy. After demitting his office on April 30, 2012, he became the DAE-Homi Bhabha Chair Professor. He is one of the leading experts in materials science and technology His comprehensive work on physical metallurgy of zirconium alloys and radiation induced order-disorder transition in alloys are widely quoted in scientific literature. His work has also provided a basis for developing a novel fabrication schedule for the pressure tubes used in the Indian pressurised heavy water reactors (PHWRs). Dr Banerjee is a recipient of many awards and honours. These include the Indian Academy of Sciences (INSA) Young Scientist Medal (1976), National Metallurgists' Day Award (1981), Shanti Swarup Bhatnagar Prize in Engg. Sciences (1989), Materials Research Society of India (MRSI) Medal (1990), G.D. Birla Gold Medal of The Indian Institute of Metals (IIM) (1997), INSA Prize for Materials Science (2001), MRSI-Superconductivity and Materials Science Prize (2003), Indian Nuclear Society (INS) Award (2003), Padma Shri (2005), MRSI Distinguished Materials Scientist of the Year Award (2008), Indian Science Congress Association's Excellence in Science and Technology Award (2010) and MRSI-CNR Rao Prize in Advanced Materials (2011). Notable among the international awards are Acta Metallurgica Outstanding Paper Award (1984) and Alexander von Humboldt Research Award (2004). He has held senior visiting positions at the University of Sussex, Brighton, UK, Max-Planck Institut fuer Metalforschung–Institut fuer Physik, Stuttgart and Forschungszeutrum, Juelich, Germany (as Humboldt Fellow and Awardee), University of Cincinnati and the Ohio State University, USA. He is a Fellow of the Indian National Science Academy (INSA), the National Academy of Sciences India, Indian National Academy of Engineering and Third World Academy of Sciences.

**S.A. Bhardwaj** retired recently as a Distinguished Scientist and Director (Technical) of Nuclear Power Corporation of India Limited, Department of Atomic Energy. As Director (Technical) he oversaw all design engineering, procurement, construction activities of pressurised heavy water reactor based nuclear power projects and related R&D activities in India. He has also been Chairman, Nuclear Fuel Complex Board, Hyderabad. He is a Mechanical Engineer, an MTech from Indian Institute of Technology, Delhi. For a major part of his career, he has been associated with nuclear fuel design, nuclear reactor core design, fuel management, reactor control and protection design. He has been an important member of nuclear safety design and its review process. He has been representing India in various international forums including the International Atomic Energy Agency meetings, World Association of Nuclear Operators etc. He is a Fellow of Indian National Academy of Engineering.

**P. Chellapandi** is a Distinguished Scientist and Director of Reactor Design Group in Indira Gandhi Centre for Atomic Research (IGCAR), Kalpakkam. He has joined IGCAR in 1979 and since then, he is working for the design and safety engineering aspects of fast reactor components for the sodium cooled fast reactors. Dr Chellapandi has made significant contributions for the 500 MWe prototype fast breeder reactor (PFBR) since its inception in 1985. His contributions include design and technology development of reactor assembly components, detailed analysis & seismic qualification of nuclear steam supply system components and thermo-mechanical consequences of severe accidents. He has outstanding academic records in Mechanical Engineering and Gold Medal with perfect grade point average of 10 in his Master Degree in field of applied mechanics, from IIT Madras. He is the Professor of Homi Bhabha National Institute. He has guided more than 110 MTech projects and more than 300 BE Projects, all relevant to fast reactor design. He has a very impressive record of publications (~620) to his credit out of which 122 are journals. He is a Fellow of Indian National Academy of Engineering. He has received many awards. To name few, the Homi Bhabha Science and Technology

Award, Indian Nuclear Society Award and National Design Award in Mechanical Engineering, Agni Award for Excellence in Self-reliance–DRDO, DAE Group Achievement Awards for Design, Manufacture and Erection of PFBR reactor assembly components. He has been bestowed upon Distinguished Alumnus Award of the IIT Madras in 2010.

**S.C. Chetal**, former Director, Indira Gandhi Centre for Atomic Research, Kalpakkam, is a graduate in Mechanical Engineering. Since joining the Indira Gandhi Centre for Atomic Research in 1971, he has been engaged in the field of fast reactor engineering and has played a key role in Indian fast reactor programme. He has made valuable contributions to the design and technical support for operation of fast breeder test reactor and design of 500 MWe prototype fast breeder reactor (PFBR), now under advanced stage of construction by BHAVINI, Kalpakkam. His contributions to PFBR include design, research & development, materials and manufacturing technology and enhancing the capabilities of Indian industries in the manufacturing of nuclear components. He is Fellow of Indian National Academy of Engineering. He has won a number of awards and honours, notable among them include: INS Award from Indian Nuclear Society (2003), VASVIK Award (2007) and National Design Award from Institution of Engineers (India), Lifetime Achievement Award from Systems Society of India (2012).

**Rajagopala Chidambaram** got his PhD and DSc degrees from Indian Institute of Science, Bangalore. He joined the Bhabha Atomic Research Centre (BARC) in 1962 and became its Director in 1990. He was Chairman, Atomic Energy Commission and Secretary to the Govt. of India in the Department of Atomic Energy (DAE) from February 1993 to November, 2000. He is currently the Principal Scientific Adviser to the Govt. of India and Chairman of the Scientific Advisory Committee to the Cabinet. Dr Chidambaram is one of India's distinguished experimental physicists and his work has been in the fields of high pressure physics and neutron crystallography. He played a leading role in the design and execution of the 'peaceful nuclear explosion' experiment at Pokhran in 1974 and also led the DAE team which designed the nuclear devices and carried out the Pokhran tests in May 1998 in cooperation with the DRDO. He has made important contributions to many aspects of our nuclear technology. He has DSc Degrees (h.c) from 20 universities from India and abroad. He has more than 200 research publications in refereed journals and all his research work has been in India. He was one of the lead speakers in the Asian Science Camps 2009 (Japan) and 2010 (India) for Science–talented youth. He is a Fellow of all the major Science Academies in India and also of the Third World Academy of Sciences, Trieste (Italy). He was Chairman of the Board of Governors of the International Atomic Energy Agency (IAEA) during 1994-95. He is a member of the Honorary Advisory Board of the international journal Atoms for Peace. He was also a member of the Commission of Eminent Persons appointed by the Director-General, IAEA, in 2008 to prepare a report on The Role of the IAEA to 2020 and Beyond. He was awarded Padma Vibhushan in 1999.

**Surendra Gadekar** and **Sanghamitra D. Gadekar** are founding members of the NGO Anumukti. They have devoted their  lives to the study of  hazards of nuclear  radiation and based on  that to social activism.  Their involvement in antinuclear protest began in 1986 following the accident at Chernobyl. The  group, Anumukti, which is based in a small village in India called Vedchhi near the then proposed nuclear power plant at Kakrapar, organised a protest rally near the plant The Gadekars have continued with the struggle  to draw attention to potential dangers of radioactivity from nuclear sites.

**Chaitanyamoy Ganguly** is an expert in uranium, plutonium and thorium based nuclear fuels and fuel cycles. He had served BhabhaAtomic Research Centre (BARC) as Head, Radio Metallurgy Division and Nuclear Fuel Complex as Chairman and Chief Executive and retired from Department of Atomic Energy as Distinguished Scientist. He was Head, Nuclear Fuel Cycle and Materials Section, International Atomic Energy Agency and

Director, CSIR-Central Glass and Ceramic Research Institute. He did his graduation and PhD in metallurgical engineering from University of Calcutta and is a post graduate from BARC Training School. He did his pre and post doctoral research on nuclear fuels in Germany and is a Humboldt Fellow. Currently, Dr Ganguly is associated with Cameco Corporation, Canada, largest uranium fuels company in the world. He is author of more than 250 research papers and is a Fellow of INSA, IASc, INAE and INASc, and recipient of National Metallurgist Award, Tata Gold Medal, MRSI Medal, Vasvik Award, Indian Nuclear Society Award and Padmashri.

**Ravi B. Grover** graduated in mechanical engineering from Delhi College of Engineering in 1970 and joined Bhabha Atomic Research Centre (BARC) Training School to study nuclear engineering. He worked as a nuclear engineer for 25 years and specialised in reactor thermal hydraulics and safety analysis. Concurrently, he obtained a PhD from Indian Institute of Science, Bangalore in 1982. Dr Grover is a Fellow of Indian National Academy of Engineering and President, Indian Society of Heat and Mass Transfer for the period 2010-2014. Presently, he is working as Principal Adviser, DAE and is a member of Atomic Energy Commission. He is concurrently working as Director, Homi Bhabha National Institute. His work profile includes energy studies, formulation and implementation of policies related to nuclear power, further evolution of nuclear legislative framework, and human resource development. He was a part of the team of officials involved in negotiations that led to the opening up of the international civil nuclear cooperation with India. Some recent awards won by him include INS Award-2006 for Nuclear Reactor Technology, including Nuclear Safety; Dhirubhai Ambani Oration Award in 2008; Distinguished Alumnus Award–2009 by Delhi College of Engineering Alumni Association; and Distinguished Alumnus Award–2011 by Indian Institute of Science and Indian Institute of Science Alumni Association.

**Frank von Hippel**, a nuclear physicist, is a Professor of Public and International Affairs at Princeton University where, in 1975, he co-founded what is now Princeton's Program on Science and Global Security; in 1989, the journal Science & Global Security; and, in 2006, the International Panel on Fissile Materials, which he co-chairs. He has worked on policy proposals relating reducing stocks of plutonium and highly enriched uranium, including ending their production for weapons; the use of highly enriched uranium as a reactor fuel; and the separation of plutonium from spent nuclear fuel. During 1993-94, he served as Assistant Director for National Security in the White House Office of Science and Technology Policy and played a major role in developing what is now called the International Nuclear Materials Protection and Cooperation Program. In 1991, the American Institute of Physics published a collection of his articles in its "Masters of Modern Physics" series under the title Citizen Scientist.

**P.C. Kesavan** currently Emeritus Professor of Sustainability Science at Indira Gandhi National Open University (IGNOU), New Delhi and Distinguished Fellow at M.S. Swaminathan Research Foundation has held previously positions as the Homi Bhabha Chair, Executive Director, M.S. Swaminathan Research Foundation, Director, Bioscience Group of Bhabha Atomic Research Centre, Mumbai and Professor and Dean, School of Life Sciences, Jawaharlal Nehru University (JNU), Delhi. Prof Kesavan is known for elucidating basic mechanisms in radiobiology. As India's delegate to the United Nations Scientific Committee on Effects of Ionizing Radiations (INSCEAR), he has consistently maintained that the biological effects observed at moderate and high doses should not/cannot be extrapolated to estimate such effects at low doses. During the 59th Session of UNSCEAR (May 2012), he successfully prevailed upon the committee to accept that ionising radiations do not induce heritable genetic changes in the humans, unlike in mice and other organisms, together with scientific evidence. He believes that nuclear energy, in agriculture, medicine and particularly in power generation, is eco friendly and an inevitable ally in achieving sustainable development. In saying this, he combines his knowledge, experience and expertise in radiation biology, sustainability science and climate change.

**Baldev Raj** has served the Department of Atomic Energy over a 42 two year period until 2011. As Distinguished Scientist and Director, Indira Gandhi Centre of Atomic Research, Kalpakkam, he galvanised a whole community of staff, scientists and engineers for advancing several challenging technologies, especially those related to the fast breeder test reactor (FBTR) and the prototype fast breeder reactor (PFBR). Dr Raj has nurtured and grown excellent schools in nuclear materials and mechanics, non-destructive evaluation, corrosion, welding, separation sciences & technology and robotics & automation. He is currently President, International Institute of Welding; President, Indian National Academy of Engineering and President-Research PSG Institutions, Coimbatore. He is also a Fellow of Indian National Science Academy, Indian Academy of Sciences, National Academy of Sciences, India and Indian National Academy of Engineering, The Third World Academy of Sciences, German Academy of Sciences, International Nuclear Energy Academy and Academia NDT, International. He is Hon. Fellow, International Medical Sciences Academy. Author of more than 850 publications in refereed journals and books, 60 books, including special journal volumes, contributions to encyclopaedia and handbooks, as well as owner of 21 patents, he has been recognised by way of more than 100 awards, 350 honours, plenary and keynote talks, editorial positions and assignments in esteemed national and international for a in more than 30 countries. He has been conferred Distinguished Alumnus Award of Indian Institute of Science, Distinguished Materials Scientist Award of Materials Research Society of India, National Metallurgist Award of Ministry of Steel, Government of India, Presidential honour Padma Shri, Indian Nuclear Society, Life Time Achievement Award (2011), Homi J. Bhabha Gold Medal Award from the Prime Minister of India during 99th Indian Science Congress (2012) and Nayudamma Memorial Award, 2012.

**R. Ramachandran** did his Graduation (B.Sc (Hons)) in 1970 and Post-graduation (MSc) 1972 in Physics from Delhi University. He obtained his Doctoral Degree (PhD) from Bombay University in 1982. His thesis work in Theoretical High Energy Physics is from the Tata Institute of Fundamental Research (TIFR), Mumbai. After that he did a short stint of teaching Physics in an undergraduate college in Mumbai between July 1981 and May 1982. Since 1984, he has been a science journalist with various publications. These include being the Science Correspondent for The Hindu and the fortnightly magazine Frontline (from July 1984 to June 1990); the Science Editor of The Economic Times (from July 1990 to September 1997); the Science Editor of the fortnightly magazine Business India (from October 1997 to October 1998). From November 1998 till date he has been with the magazine Frontline, initially as its Science Correspondent (till 2009), later (till June 2011) as an Associate Editor and at present a Consultant. All along he has also been writing for the daily The Hindu.

# Foreword

The Indian National Science Academy (INSA) has a key role in promoting science and technology in the country through recognition of excellence achieved by Indian scientists and engineers and undertaking multifaceted activities to achieve this goal. It also represents India on behalf of the Government of India in various activities of the International Council for Science (ICSU) and its various unions. The Academy has established bilateral collaboration with nearly 50 academies all over the world and is strongly linked with several global networks of academies and science forums. In addition, the Academy regularly organises workshops/symposia/ seminars on issues of topical value in science and of societal concerns. During the last one year, the following two meetings were held:

i)   Man, Animal and Science—September 2011; and

ii)  Challenges in Nuclear Safety—February 2012.

The inherent spirit in the deliberations in these workshops was to discuss and debate on different points of views on the subject, some of which may be opposing each other and final conclusions may be drawn based on scientific evidence.

The Workshop on Challenges in Nuclear Safety was organised during February 14-15, 2012, keeping in view the deep general concern about the safety of Indian nuclear installations in the backdrop of the Fukushima catastrophe in Japan. The main topics discussed in this meeting were: nuclear energy options for the country; safety measures that have been taken to ensure nuclear reactor safety and address radiation concerns; impact on the population or surrounding vegetation and lives; the science and technology of fast spectrum reactors and closed fuel cycles for nuclear power generation in future;

and results of studies on radiation health hazards, impact on aquatic eco systems, bio-diversity and agriculture in the vicinity of nuclear power plants. The issues of economics of nuclear energy and the legal provisions for safety and liability and national framework for governance of nuclear powers, were also discussed. Besides senior scientists and responsible officers from Department of Atomic Energy, several representatives of different organisations including Swaminathan Research Foundation, Chennai; Nuclear Power Corporation of India Ltd. (NPCIL), Mumbai; Editor of Anumukti, and a senior journalist participated in the deliberations and contributed written manuscripts on their respective topics. The Academy was very glad that Prof Frank von Hippel from Princeton University, USA, who is Co-Chair of International Panel of Fissile Materials, actively participated in the deliberations and he enlightened the audience about history and current status of closed fuel cycle and breeder reactor programmes worldwide. It was heartening that top some of the top scientists of the country in this field, particularly Dr R. Chidambaram, Principal Scientific Advisor to Government of India, Dr S. Banerjee, the then Chairman, Atomic Energy Commission (AEC) and Secretary Department of Atomic Energy (DAE), Dr R.K. Sinha, the present Chairman, AEC and Secretary DAE, Dr Baldev Raj, President, INAE and former Director Indira Gandhi Centre for Atomic Research (IGCAR), Dr R.B. Grover and Prof P.C. Kesavan contributed significantly.

Prof R. Rajaraman, Vice President, INSA together with Dr Seema Mandal and other staff of Science and Society Section of INSA, worked hard in organising the Workshop. It is through the sustained efforts of Prof Rajaraman, this document could be prepared.

I would like to put on record on behalf of the Academy and on my own behalf the systematic and high quality efforts made by Prof Rajaraman with the assistance of Dr Seema Mandal and other colleagues from INSA. It is our hope and wish that this book will serve as a source of important information covering different aspects of nuclear safety for a wide variety of leadership.

—Krishan Lal
*President*
Indian National Science Academy, New Delhi

# Preface

An important component of India's science and technology policy and one that has been generating considerable public debate in recent years, is nuclear energy. Many factors contribute to its prominence.

To start with, maintaining the current rate of growth of our economy requires a correspondingly rapid increase in energy production. If the country is able to restore its earlier 9 per cent growth rate for the GDP, it will have to be accompanied by the capacity to produce about 600 Gigawatts of electricity (GWe) from all sources by 2030, as against the roughly 200 GWe available today. A somewhat slower GDP growth may require less, but would still set a staggering target to reach by 2030. Such generating capacity is needed not just for our industries, but also for meeting the energy needs of the rural and urban poor. Depending on how you define electrification, anywhere up to about 100 million villages are still not electrified in our country.

Meanwhile, the international community is deeply concerned about the effects of carbon emissions from coal and hydrocarbon fuels. This has led to an increasing interest in generating electricity from non-carbon emitting sources, of which nuclear is a leading candidate. It has a relatively small carbon footprint and in terms of displacement of people from the site, occupies a smaller area per Megawatts of electricity (MWe) than solar, wind, hydroelectric or any other source.

If nuclear energy is to make even a 10 per cent contribution to our projected electricity requirement of 600 Gwe in 2030, this is going to call for a capacity of 60 GWe—a huge increase from the

4.78 GWe, which is currently being generated by the nuclear sector. That is a very challenging goal.

In this context, two major developments have occurred in the recent past, one of them favorable to India's nuclear energy programme, and the other less so. The first was the Indo-US Nuclear Deal achieved in 2008, after three years of hard negotiations by the Indian government, not only with the US and the Nuclear Suppliers Group (NSG) of countries but also with domestic critics. The deal helped lift longstanding international nuclear sanctions imposed against India. We can now buy enough uranium to fuel a large nuclear expansion, our own reserves of this metal being limited. We can also enter into collaboration with reactor builders from other nations to speed up the expansion. After the consummation of the Nuclear Deal, negotiations have been going on with reactor builders from France, the US and Russia.

The second development—this one negative—has been the growing concern about, and public resistance to setting up of more reactors. Part of this resistance stems from the 'development versus displacement' dilemma common to any large industrial initiative, resulting from the displacement of the populace from the site of the facility, and the disruption and possible loss of their traditional livelihoods. Another part is special to nuclear reactors, *viz.*, the fear of radiation hazards, not just in the event of a nuclear explosion but also from the daily proximity of nuclear reactors. This last concern has been vastly enhanced by the recent reactor explosions at Fukushima. In recent months, these concerns have led to sizeable public protests especially at Jaitapur and Kudangulam, where new reactors are being planned and built.

There were two other offshoots of public concerns about the hazards posed by nuclear reactors. One was the Nuclear Liability Bill and the other a pending bill for an independent nuclear regulatory authority. The Liability Bill was passed into law by the Indian parliament in 2010. Coming as it did at the same time as the widely observed 25th anniversary of the Bhopal gas leak tragedy, the lawmakers were under pressure to ensure that in the event of a similar nuclear tragedy, foreign builders of our reactors would be held liable for compensation, if the damage could be traced to manufacturing defects. As a result, we got a Liability Bill more stringent in some ways than other nations. India is the first country to demand supplier liability for nuclear reactors. The Bill was also people friendly in setting a total sum of ₹ 1,500 crore for the compensation (more for instance than China), along with the requirement that the reactor operators (in our case, the government) disburse those funds to affected persons within 3 months, without waiting for elaborate judicial procedures.

The other legal instrument that attracted attention as a result of Fukushima, concerns the nuclear regulatory authority. Up until now, the body vested with this regulatory function was the Atomic Energy Regulatory Board. It was, however, a wing of the Department of Atomic Energy (DAE), that got its funds from the latter and reported to its Head. This was clearly not a satisfactory situation, with the regulator being under the authority of the regulated. The government has accepted the need to correct this situation and a new Nuclear Safety Regulatory Authority (NSRA) Bill has been sent to the parliament proposing a more independent regulatory authority.

Unfortunately, we couldn't arrange a chapter on the Liability Bill, but we do have a detailed discussion of the Regulatory Authority bill currently before the parliament at the time of writing.

In addition to these societal concerns about developing nuclear energy, there are also major technical issues at the core of our nuclear programme on which there are conflicting views among international experts. These include the pros and cons of a closed fuel cycle, the thorium programme, the special difficulties associated with fast breeder reactors, reprocessing technology, the costs and feasibility of safe disposal of spent fuel and decommissioned reactors. Not only is it appropriate for INSA, as an apex scientific body, to discuss these technical matters, it is indeed vital that their deliberations be brought to the attention of those who make or influence policy in the energy sector in a volume such as this. The ultimate success of India's ambitious drive to expand its nuclear energy sector depends crucially on the choice of the technologies adopted. The DAE's long term flagship plan has been based on Dr Bhabha's 3-stage programme, which aims to use India's large reserves of thorium ore by adopting precisely the technologies mentioned above, on which international experience has been mixed.

The aim of this collection of articles and of the INSA Symposium on which it was based, is to discuss these issues in some detail. What, in our view, makes our exercise very unique is that for the first time a mix of the top leadership of the government's nuclear programme on the one hand, and a set of non-governmental expert critics with serious concerns about the programme on the other, have amicably participated in both.

One might have thought that in 60 odd years that our nuclear programme has been on, there would have been many such symposia representing all viewpoints on its pros and cons, since such debates are clearly healthy and essential for providing checks and balances in any democracy. Unfortunately, however, meaningful interactions between proponents and critics of the nuclear programme have rarely taken place in the past.

Primarily, this was not the fault of either side, but a consequence of how the nuclear programme in India was established and how it evolved.

Opponents attribute the absence of such debate to unwillingness on the part of the nuclear establishment to engage in serious discussion with the public or with non-governmental scientists. To some extent this may have been true in the past, and that should not be too surprising. Few large governmental organisations go out of their way to be more transparent than is necessary for their functioning. The DAE, set up with a total monopoly in the nuclear field, generous governmental funding and wide support across the political spectrum, was particularly under little pressure to do so. On top of that, having to develop complex new technology under the handicap of international nuclear sanctions, may have lead to closing of ranks and a siege-like mentality. Lastly, the fact that the same DAE was also engaged in developing nuclear weapon capability, an unavoidably secret activity, contributed further to opacity.

Having said that, it must be acknowledged that as far as the production of this volume of articles and the organisation of the INSA seminar was concerned, DAE has been most supportive. They readily agreed to engage with any responsible critics in a non-confrontational discussion and offered their most distinguished scientists as speakers. Despite the busy schedules that go with apex organisational positions, their speakers also took the trouble to submit their contributions to this volume on time.

Turning to critics of the nuclear energy programme, they too were, by and large, unwilling to engage in calm and serious discussions despite their sincerity and commitment to their concerns *viz.*, reactor safety, hazards of radioactivity and other environmental effects. These are very valid issues to raise, but they all necessarily involve fairly complex science and technology in which, with notable exceptions, most critics were themselves not experts. They based much of their criticism on anecdotal evidence and hearsay laced with emotional and ideological arguments, rather than on first hand technical knowledge or comprehensive empirical data. Some of them adopted the methods of social activism over detailed substantive discussions. As a result, in place of a meaningful dialogue, each side has ended up preaching to its own choir.

Yet, anti-nuclear activists and writers can also not be entirely blamed for their approach. They too had little choice. The fact is that almost all the technical expertise in India in nuclear technology lies inside the walls of the DAE. This happened by design from the start, presumably to conserve manpower in this new technology. The DAE set up its own training school whose graduates were absorbed into its various emerging nuclear facilities. Outside of this, there was no programme of graduate courses in nuclear technology in universities, or with one small exception, in the IITs. As this pattern continued for decades, there developed a skewed situation with a mass of high level nuclear expertise within the DAE and practically none outside.

The net result has been that strongly conflicting views in the public domain on the safety and desirability of nuclear energy came no closer to resolution than before, leaving the larger intelligentsia (let alone the lay public) unable to make an educated decision on this very crucial component of India's energy policy.

But, in the past decade things have improved on this front. Gradually more and more people in India, outside the atomic energy establishment have picked up some expertise in one or the other aspect of nuclear technology, its environmental and radiation effects, its economics, legal structure etc. The three year long public debate on the Indo-US Nuclear Deal also spurred commentators from academia, the media and the political class to learn about Plutonium, Uranium, Breeder reactors and so on. The number of 'outside experts' is still not large, but was large enough to allow us to gather together an excellent set of expert contributors for this volume.

The range of topics covered in book is such that there is some inhomogeneity in style and content between the different chapters. Some, dealing with the overall policy matters, future plans for the nuclear programme, legal and regulatory instruments etc., are discussed in plain English prose. Other chapters, which discuss radiation hazards and environmental issues or the choice of technology involving with a closed fuel cycle are presented in considerable scientific detail, to the point where they could well belong to a standard scientific journal.

While such lack of uniformity within the same volume may violate editorial aesthetics, we feel it is unavoidable here if we are to achieve our main goal—to address issues of public policy and concern in some definitive fashion, using whatever style and form of discussion each issue permits. For instance, while concerns about reactor safety and health hazards can be stated in plain language, their resolution necessarily involves going into considerable scientific detail. Otherwise, the debate reduces to just conflicting assertions from each side, where neither party convinces the other. We do not intend that every reader should wade through all the scientific portions. Most may be content with the introductory and concluding sections. But, for those who would like to come to their own conclusions based on hard information, much of the required material is given right there in the chapters, with full references to the literature.

Last but not least, this entire initiative would not have been possible but for the kind encouragement of Dr Krishan Lal, the President of INSA. Equally important was the support of the DAE leadership, particularly Dr R. Chidambaram and Dr S. Banerjee, both former Chairmen of the Atomic Energy Commission (CAEC), and Dr Ravi Grover, member, AEC.

I am also grateful for the unstinting cooperation of officials of the INSA Secretariat, particularly our Executive Secretary Dr Alok Moitra and Dr Seema Mandal.

—R. Rajaraman
*Emeritus Professor,*
Jawaharlal Nehru University, New Delhi
and *Vice President,*
Indian National Science Academy (INSA), New Delhi

# I

# Overview

1

R. CHIDAMBARAM

# The Nuclear Energy Option and Nuclear Safety

## Introduction

A large country like India has enormous energy needs and all energy options are important for India. These options have also to be evaluated from the point of view of sustainable energy security and of mitigation of the global climate change threat. Viewed in this context, the significance of the three-stage Indian nuclear programme—pressurised heavy water reactors using natural uranium in the first stage, to which can be added imported light water reactors; fast breeder reactors using plutonium as fuel and thorium as blanket in the second stage; and uranium-233-thorium cycle in the third stage—becomes clear. For sustainable energy security and the optimum utilisation of the uranium resources in the world, the nuclear fuel cycle has to be closed.

The excellent safety culture of the Department of Atomic Energy—this has been carefully nurtured since the inception of the atomic energy programme in the 50s—has to be projected more widely; for example, the fact that the Bhabha Atomic Research Centre has been at the forefront of global research in the area of nuclear waste management. It is also apparent—particularly after the uninformed recent protests against the Kudankulam project, which have delayed the project and prevented earlier availability of much-needed electricity to the southern region for societal needs and for industrial production—which neighbourhood public awareness programmes by the nuclear community have to be strengthened.

I must also mention in passing that there is some semantic confusion in the media among the phrases 'nuclear safety', 'physical security of nuclear weapon-usable materials' and 'nuclear safeguards'. This is particularly true for Indian languages; for example, the word *nabhikiya suraksha* is often used in Hindi to translate all the above three words!

## The Human Development Index

While the United Nations uses main three parameters—per capita GDP, life expectancy at birth and adult literacy—to calculate the Human Development Index (HDI), I have been saying for more than two decades (see, for example, Chidambaram, 2001) that, a developing country like India needs only two parameters—per capita electricity consumption and female literacy to obtain the HDI value. I prefer female literacy to adult literacy because it is a measure not only of literacy but also of equity and justice in the society. Female literacy correlates strongly and inversely also with infant mortality and birth rate. I have also shown that, while per capita electricity consumption (PCEC) is (obviously) monotonically related to per capita GDP, it also correlates very strongly with health parameters, including the ultimate health parameter, life expectancy at birth. If we plot HDI against PCEC, we obtain an S-shaped curve (see Figure 1.1). It is flat at the bottom because, even in a very poor country or in a very poor segment of any society, a human being will survive for about forty years; at the other end, the human system begins to break down at about eighty years, even if the standard of living is very high. In between the flat portions, there is a steep portion, which India has been climbing rapidly since independence. From where we are now to reach at least the upper cusp of the HDI *versus* PCEC curve, i.e., the India of our dreams, the per capita electricity consumption in India must go up by about six times.

**Figure 1.1**

*Variation of Human Development Index (HDI) with Per Capita
Electricity Consumption (PCEC)*

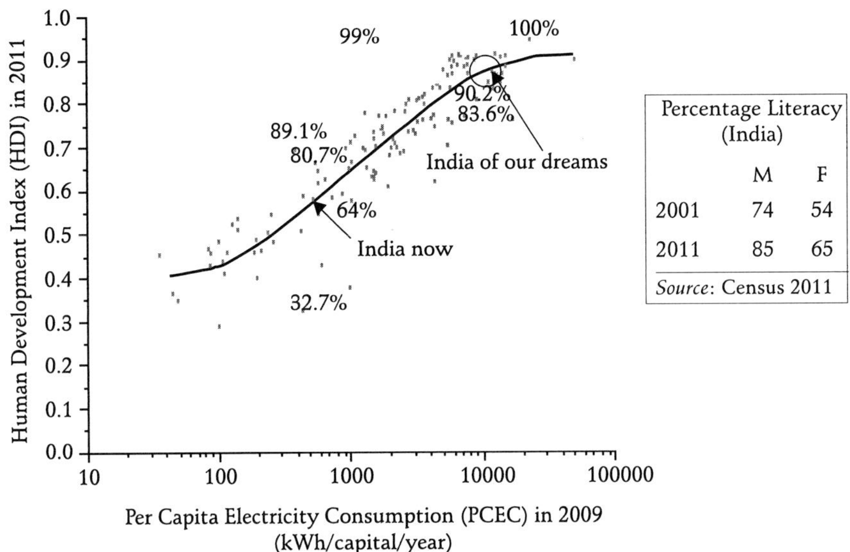


*Note*: Each point represents one country. If the female literacy is high, the country goes
above the average curve; and if the female literacy is low, the country goes below the
average curve

*Source*: World Bank (2011). *Human Development Report, 2011*; CIA, *World Factbook*.

## Energy Sources

That is why all sources of energy are important for India—
fossil fuels, particularly coal; renewable energy sources (hydro,
solar, wind, etc.) and nuclear. The future possibilities include
biofuels, shale gas, nuclear fusion, etc. It is very difficult to project
the future global comparative roles of different sources of energy.
It will depend on starting assumptions, of course, but also in
any country on locally available fuel resources and its access to
technologies. Recently, Sukhatme (2011) said: "The estimates made
here indicate that even with a frugal per capita electricity need of
2000 KWh/annum and a stabilised population of 1700 million by
2070, India would need to generate 3400 TWh/yr. As opposed to
this, a systematic analysis of the information available on all the

renewable energy sources indicates that the total potential is only around 1229 TWh/yr (438 TWh/yr from Solar)…It is concluded that, in the future, as fossil fuels are exhausted, renewable sources alone will not suffice for meeting India's needs." My own estimate of per capita electricity need, before India becomes a 'developed country' in the fullest sense of the term (a time when the quality of life in rural India becomes comparable to the quality of life in the non-urban areas of already developed countries), is three times higher than the above 'frugal' estimate of Sukhatme.

## The Nuclear Energy Option

In late 2007, the Director General of the International Atomic Energy Agency (IAEA), Dr Mohamed ElBaradei, appointed an independent Commission "to advise on how the nuclear future might evolve to 2020 and beyond, what the world is likely to demand of the IAEA, and what steps need to be taken to allow the IAEA to fill those needs." In the report[1] submitted in May 2008, the Commission said *inter alia* that: "Expanded use of nuclear technologies offers immense potential to meet important development needs. In fact, to satisfy energy demands and to mitigate the threat of climate change—two of the 21st century's greatest challenges—there are major opportunities for expansion of nuclear energy in those countries that choose to have it."

Lessons have been learnt from the recent unfortunate Fukushima accident, particularly on the continued functioning of post-shutdown cooling systems after even remotely possible extreme natural events. Such safety reviews have been carried out in all leading nuclear countries in the world including India. But the above conclusion of the Commission on the importance of the nuclear energy option remains unchanged.

In the two decades between 1970 and 1990, the nuclear power growth in the US and in France was very rapid, the former growing to 96,000 MWe and the latter to 56,000 MWe. This shows that

---

1. Report prepared by an independent Commission at the request of the Director-General of the International Atomic Energy Agency, "Reinforcing the Global Nuclear Order for Peace and Prosperity: The Role of the IAEA to 2020 and Beyond", May, 2008. This author was a member of this Commission.

such rapid growth of nuclear power is possible if a country has the technological capability. Then, the growth slowed down and the current net nuclear power is 101,000 MWe in USA and 63,000 MWe in France. This was partly because of the Chernobyl accident in 1996, but we must remember that developed countries at the top of the HDI *versus* PCEC curve already have a very high level of PCEC (over 75% of the electricity in France comes from nuclear power), they have begun to learn the merits of introducing energy efficiency because of the climate change threat, and can afford to ignore new nuclear energy inputs in the short term.

On the other hand, large energy-stressed developing countries like India and China desperately need nuclear power inputs for energy security. China started their nuclear power programme two decades after us and their current nuclear power capacity is double to that of ours. South Korea has rapidly increased its nuclear capacity to nearly four times of ours and sees itself now as an exporter of nuclear power plants. The Indian nuclear power programme, however, is based on a strong indigenous research and development (R&D) foundation, it is one of the few countries in the world with comprehensive capabilities in the full nuclear fuel cycle, and its emphasis on the closed nuclear fuel cycle has important relevance to the global nuclear community. We have shown (Chidambaram *et al.*, 2007) that if nuclear is to be a sustainable mitigation technology in the context of the climate change threat, the nuclear fuel cycle has to be closed.

## Nuclear Safety in Indian Reactors

The two Light Water Reactors nearing completion at Kudankulam are based on an advanced model of the Russian Pressurised Water Reactor (PWR) VVER-1000 that have state-of-the-art safety features, and have been comprehensively reviewed by a task force of the Nuclear Power Corporation of India Ltd (NPCIL). These safety features include hermetically-sealed double containment, passive heat removal system, redundant safety systems, additional shut-down systems, core-catcher system, etc. It also has state-of-art instrumentation systems. After careful analysis, it has been found that there is no tsunami risk to the reactors. I must add that there are 16 Kudankulam type of reactors operating in Russia and nine in other countries. In particular, two VVER-1000 type of reactors of the Kudankulam type have been operating well in China Tianwan Phase-I. Both of them went into commercial operation in 2007 and six more such units are planned in Tianwan. Late last year, Belarus and Bangladesh signed contracts with Russia for the construction of their first nuclear power plants, based on the VVER pressurised water reactors. For the European Pressurised Reactors (EPRs), planned for Jaitapur also, it has been found that there is no tsunami risk—the nearest tsunamigenic seismic source is about 1400 km away for Jaitapur, compared to the distance of 130 km for the tsunamigenic fault in the case of Fukushima. The plant grade is also planned at 10 metres above mean sea level, with a built-in margin of 2.5 metres and additional tsunami/storm surge margin of 3 metres as per codal requirements.

After the Fukushima accident, as directed by the Prime Minister, the Department of Atomic Energy (DAE) and the NPCIL constituted six task forces to review the consequences of similar situations in the Indian Nuclear Power Plants (NPPs). The main findings of the task forces are: continued decay heat removal mechanisms exist in Indian NPPs, with inventory augmentation through isolation condenser for the Boiling Water Reactors, Tarapur Atomic Power Station (TAPS) 1&2 at Tarapur and with steam generators for the PHWRs (with the provision of further inventory augmentation through fire water system, etc). These safety reviews found that all Indian NPPs are meeting current safety standards through regulatory review by the Atomic Energy Regulatory Board (AERB) and that all the AERB recommendations, based on the Three Mile Island and Chernobyl accidents, are in place.

Safety is in design and in operation; safety has to be strengthened through regulation. But, most importantly, safety is assured through the existence of a 'safety culture'. In more than four decades of our operation of nuclear power plants, our excellent track record in safety is because of the safety culture in the Department of Atomic Energy.

## Conclusion

The support to nuclear energy continues in most countries, particularly in energy-stressed developing countries. The World Nuclear Association (2012) in its recent news report said, "Support to nuclear energy in the United Kingdom has returned to levels seen a few months prior to the Fukushima accident in Japan, a new public opinion poll indicates…the anti-nuclear feeling following the March 2011 accident now looks like no more than a temporary blip, as year-on-year improvement in support has resumed." In the IAEA General Conference in Vienna in September 2011, six months after Fukushima accident, the Chinese representative quoted an old Chinese proverb: "Should not stop eating for fear of choking!" What he meant was that, while safety reviews should be there, the response should not be to reject the essential nuclear energy option.

It is important to communicate, even more than before, to the people living in the neighbourhood of our nuclear power plants facts about the importance of the nuclear energy option to the country and to their own region and the safety features of our nuclear power plants (existing and planned), and how the communities around the nuclear power plants have greatly benefited after their establishment. Tarapur was a quiet fishing village in Maharashtra before the nuclear plants started coming up, now its economy has drastically improved. Our communication to the public must be correct, concise and comprehensible. The DAE is also planning out-reach programmes to make available relevant technological knowledge to the rural communities around nuclear power plants, through proximate academic institutions. The National Knowledge Network, which has already connected all the DAE institutions, will be very useful in this context.

The knowledge from the nuclear community is often helpful also in pursuing non-nuclear technologies. The Commission report I referred to earlier[2] has said: "Within the energy sector, the nuclear community around the world, which the IAEA is uniquely positioned to network, can contribute significantly to other technologies. In renewable energy technologies, for example, the

---

2. See note 1.

nuclear community's extensive knowledge is a valuable resource in areas such as thermal engineering, materials and computational fluid dynamics." Indira Gandhi Centre for Atomic Research (IGCAR) is a partner in the R&D programme towards an 800 MWe Advanced Ultra-Supercritical Thermal Plant, along with the Bharat Heavy Electricals Ltd. and the National Thermal Power Corporation. IGCAR's strength is in materials development, because of its long experience in the development of fast breeder reactors.

Before concluding, I must congratulate the President of the Academy, Dr Krishnan Lal, and Vice-President, Prof R. Rajaraman, for this important initiative in organising a seminar on nuclear energy. INSA must play a pro-active role in bringing together scientists, technologists and fellows of the Academy to discuss in depth issues of public interest which have a strong science and technology content. This is one example.

## References


Chidambaram, R. (2001). *Current Science* 81(1): 17-21.

Chidambaram, R., R.K. Sinha and A. Patwardhan (2007). *Nuclear Energy Review*: 38-39

Sukhatme, S.P. (2011). *Current Science* 101(5): 624-30.

World Nuclear Association News, London, 18 January 2012.

2

S. BANERJEE

# Nuclear Power

## *Some Frequently Asked Questions*

After 25 years of the Chernobyl accident when the prospects of a nuclear resurgence were imminent, the accident in Fukushima raised a fresh debate on nuclear power. R. Chidambaram in his chapter has outlined the programme of Department of Atomic Energy and its expected impact on Indian energy scenario. Let me now highlight the major concerns, which have been raised in the media and in several debates in the recent times on the growth of nuclear power in India. I have listed them in the following:

- Why nuclear energy?

- Can renewable energy resources meet the entire demands of electricity in India?

- Why some countries are abandoning nuclear energy?

- Is the environment around nuclear power plants and fuel cycle facilities excessively radioactive?

- What is the necessity of building some Light Water Reactors with international cooperation in addition to building more indigenous pressurised heavy water reactors?

- How long it will take for us to reach the third stage of our three stage programme and start tapping energy from thorium?

- Is the regulatory mechanism prevailing in India, sound?

- How to ensure safety of imported reactor technologies that are being considered?

- What are the key elements of the Civil Liability of Nuclear Damage Act?

- What are the permanent solutions for high level radioactive waste?

The growth of electricity demands in India has been worked out by Planning Commission on the basis of 8 or 9 per cent growth in economy and it has been estimated that the installed capacity will touch a figure of 800 to 1000 GWe by 2030. DAE has made a more modest projection, which shows that by the year 2030 or so, the installed capacity of electricity generation in the country will reach a value of about 600 GWe. It is imperative that for meeting such extremely large-scale power production, one has to make use of all the proven sources of energy generation. One cannot plan to meet India's vast energy requirements by deploying just one or two sources of energy generation. There are estimates made for the growth in power generation from renewables, including hydro-electricity. These projections also point out that the gap, between the total demand and what is possible to generate from renewable, will be about 400 GWe by the year 2030. Now, this deficit needs to be covered by increasing the capacity through thermal and nuclear power generation. If this is done entirely by thermal power, we have to burn close to 2 billion tons per year of coal, which will generate about 3Gte of $CO_2$ emissions. In order to reduce generation of such a quantum of $CO_2$, which will have a serious impact on the climate, the share of nuclear power needs to be enhanced in the next few decades.

Recently, Sukhatme (2011) has made a realistic evaluation of energy availability from all renewables and has shown that a substantial deficit will exist, which needs to be covered by additions in nuclear and thermal power generation. With rapidly depleting fossil fuel resources, the share of nuclear energy needs to grow continuously for supplementing energy supply. Our dependence on the two primary energy sources—solar and nuclear—will continue to rise in the long term future. I believe this argument explains in a nutshell, why we need nuclear energy both in the near and the distant future as only the renewable energy resources will not be adequate to meet the electricity demand in India. It is expected that the talks, which will follow in the present seminar will provide quantitative data in justification for the need of nuclear energy in India.

Often we hear that some technologically advanced countries are abandoning nuclear energy. Let us take the example of Germany, which has proclaimed that by 2022, they will not produce nuclear energy in the country. I looked at the growth in electricity generation in Germany during the last five years and I have found that the electricity generation has declined by about 1 per cent during the said period, while that in India has grown by over 50 per cent. We have to bear in mind that even today, a large section of our population, nearly 40 per cent, do not have access to electricity. Moreover, for a country such as Germany can import electricity from neighbouring countries such as France and Czech Republic, where the share of nuclear power is very large. One should also keep in mind that the population in Germany has not only stabilised but also has slightly declined in the immediate past, whereas in India a stabilised population is expected only after about four decades.

Several concerns have been expressed regarding radiation environment around nuclear power plants and fuel cycle facilities. In this respect, I would like to bring the following facts to your attention. The world's average of natural background radiation, a major part of which coming from the cosmic radiation is about 2400 microsievert per year (mSv/y). Background radiation varies considerably from place to place, for example in Mumbai, the background level is about 50 nanosievert per hour (nSv/h), while in Hyderabad it is about 260 nSv/h. If one plots these variations from place to place, it can easily be seen that the location, where nuclear power stations and fuel cycle facilities are present, the levels of radiation lie within the scatter band of background from one place to another, the average additional dose received from nuclear power plants by the public right at the boundary of the exclusion zone varying between 1 to 20 mSv/y. This is an insignificantly small fraction, not only of the average background dose (2400 mSv/y) but also of the maximum Atomic Energy Regulatory Board (AERB) permissible limit of 1000 mSv/y. There is yet another way of looking at these data. While the average radiation, over and above the natural background value, received by the public living close to nuclear power plants is less than 10 mSv/y, a single chest X-ray examination or a to and fro air trip between Mumbai and Tiruvanathapuram will give about 50 mSv each time. A single chest CT Scan can give dose of nearly 7000 mSv each time. It is clear from these numbers that the radiation dose received from nuclear power plants or from fuel cycle activities is an extremely small fraction of what we receive from the background or from other human activities.

Many are raising the question of why India is planning to go for technical cooperation with other countries when it has his own established reactor technology. The country is indeed proud that we have developed the total technology of nuclear power reactors and have mastered the entire fuel cycle. Today, 20 reactors out of which 18 are indigenous pressurised heavy water reactors (PHWRs) are in operation. They have impressive capacity factors—over 90 per cent in the reactors provided with adequate fuel supplies, a record of accident free performance for 340 reactors, in years for operations, considerably low capital cost per megawatt of installation and a competitive tariff of nuclear electricity (typical ₹ 2.50 per kilo watt hour (kwh), as against nearly ₹ 4 from thermal power station using imported coal). India is also moving ahead in its well charted three stage nuclear power programme to achieve sustainable nuclear power generation by exploiting the vast reserves of thorium in the country. In spite of all these impressive achievements we see, for meeting the immediate demands of electricity for the next 2 or 3 decades, the share of nuclear power must be enhanced rapidly, while it is also well known that the operation of our three stage programme would require building of adequate fissile inventory. The growth of fast reactors will not only require a large inventory of plutonium by reprocessing spent fuel from existing PHWRs, but also establishing capacity for large scale reprocessing of spent fuel arising from fast reactors. Reprocessing activities in the country is growing at a reasonably fast pace and the industrial scale facility for reprocessing fast reactor spent fuel is expected to come up during the next five years. The rapid capacity growth with indigenous uranium, as well as with indigenous three stage programme, will therefore require a few decades. In the intervening period, there is a necessity for adding nuclear power plants of large capacities using imported uranium. The opening up of civil nuclear cooperation with several countries will make it possible to implement this strategy.

The closed fuel cycle adopted by India essentially implies that the resource base of nuclear fuel is enhanced many fold by converting fertile material into fissile. Energy from uranium does not remain confined to the naturally occurring 0.7 per cent of fissile U-235 nuclide, we can exploit the fertile component of 99.3 per cent of U-238 through their conversion into plutonium, which is subsequently used as fuel in the fast reactors. Excess neutron available in fast reactors is again used for conversion of thorium into U-233, which can fuel the reactors in the third stage. This multiplication process can as well be employed for the imported uranium and the full potential of energy stored in the fissile plus fertile material can be exploited following our three stage programme. Setting up of a few light water reactors (LWRs) with international cooperation in the intervening period for enhancing nuclear capacity will therefore give a big fillip in our plan to extend nuclear capacity employing fast reactors and sustaining that capacity for a long period with thorium fuel reactors in the third stage.

It is difficult to make precise prediction regarding the time frame for a large scale utilisation of thorium for power generation. This will depend on how fast we are growing our installed capacity, reprocessing capacity and developing technologies associated with thorium fuel cycle. In each of these areas, focused activities are ongoing and we are confident in implementing the programme as is originally envisioned.

It is of is no doubt that there is a prerequisite of a robust nuclear safety regulatory mechanism for establishing a nuclear capacity, which will be a significant part of the overall energy mix. The regulatory process, which is now being monitored by AERB is indeed technologically sound and is fully effective. The teeth that AERB gives them is the authority for even closing down an activity, which has not been complied with an AERB stipulation. Recently introduced Nuclear Safety Regulatory Authority (NSRA) bill has right provisions for making the authority a statutory body fully empowered for effective safety regulatory control of nuclear installations.

India has made it very clear to the prospective technological partners of other countries that the full scale safety evaluation will be carried out by Indian regulatory authorities, before they are accepted for installation in this country. This is in addition to the requirements that the technology must be cleared by the regulatory authority of the country of origin. If you take the example of VVER reactors, which are being setup at Kudankulam, this process was meticulously followed and safety clearances were obtained from AERB at every necessary stage.

There have been question on who takes civil liability in case of occurrence of a nuclear event, which affects the public or the environment. The Civil Liability of Nuclear Damage (CLND) Act, which has been passed by the parliament makes it amply clear that the liability rests on operator to the limit of up to ₹ 1500 crore beyond which the government takes the liability. India has signed an agreement to be a part of the International Convention of Supplementary Compensation and once this convention is ratified, India can claim compensation from this International Convention beyond ₹ 1500 crore and up to a limit of 300 million Special Drawing Rights. If the damage is beyond this limit, the government is going to take the liability. Perhaps the most important element of the CLND Act is that this stipulates prompt settlement of the compensation claims and clearly identifies the agencies who are liable. Since this liability is a no fault liability, the victims need not have to establish cause-effect relationship for making their claims. In contrast, operators have a right of recourse to make the claims from the suppliers, provided they can establish the fact that the fault of the supplier has resulted in the nuclear damage.

There is an understandable concern about high level radioactive waste. However, the volume of high level waste is relatively very low and it is technologically amenable for safe long-term storage.

Some of the radio isotopes present in the spent nuclear fuel have radioactive life of about 20000 years. This makes the total radio-toxicity inventory of the spent fuel above that of natural uranium ores for such a long period. However, if plutonium, americium and some minor actinides are removed, the radiotoxicity inventory will fall below that of natural uranium ore in about 300 years of time. Closed fuel cycles have the advantage of being able to burn the long lived radioisotopes and reduce the spent fuel radio toxicity inventory to a much shorter period. Furthermore, technological solution to the long-lived waste issue is expected to be available from the development of high energy, high current, ion accelerators, which will transmute long-lived radioisotopes to significantly shorter radioactive life.

Lastly, let me thank the Indian National Science Academy (INSA), Prof Krishan Lal, President, and Prof R. Rajaraman, Convener, who have organised this important meeting on nuclear power. After 25 years of the Chernobyl accident when the prospects of a nuclear resurgence were imminent, the accident in Fukushima raised a fresh debate on nuclear power. Against this backdrop INSA has taken a bold step in conducting a seminar-cum-debate on several aspects of nuclear power. Undoubtedly, this is a very timely and welcome step for addressing several queries which are coming in the minds of people.

## References

Sukhatme, S.P. (2011). *Current Sciences* 101(5): 624-30.

# II

# Reactor Safety and Radiological Concerns

# 3

S.A. BHARDWAJ

# Nuclear Reactor Safety

## Introduction

The fission reaction caused in uranium-235 nuclei through interaction with neutron, converts the mass defect in the nucleus to a large amount of energy, about a million times more than that obtained from combustion of carbon atom. Thus nuclear reactors need relatively much less quantity of fuel for producing same power when compared with fossil fuel. The fission reaction breaks the uranium nuclei into two smaller nuclei and along with the release of huge amount of energy, it also produces two or three free neutrons. One amongst these free neutrons is necessary to sequentially cause fission reaction in other uranium-235 nuclei to maintain the equilibrium fission 'chain reaction' and generate energy at a constant level. Having met this requirement of maintaining the chain reaction, certain excess of free neutrons remains available. Some of these may leak out from the system or get absorbed in structural materials present in the reactor. However, by appropriate physics design of the reactor, it is possible to absorb a significant fraction of these excess neutrons in uranium-238, which is an abundant fraction (99.3%) of natural uranium. Such absorption of neutron by uranium-238 creates plutonium-239—a manmade fissile isotope—a nuclear reactor fuel. A similar absorption of such spare neutrons in thorium would produce fissile isotope, uranium-233.

Thus, fission reactions are capable of generating large amount of energy and simultaneously convert fertile isotopes like uranium-238 and thorium-232 into fissile isotopes to be recycled again as fuel. This natural phenomenon permits to adopt nuclear energy to meet energy demand for a very long term and ensure energy security. The fissile atoms produced in the reactor can be separated from the discharged fuel by chemical processing for reuse, again in nuclear reactor along with left over uranium-238 to produce more energy and more fissile material. This is the genesis of three stage nuclear power programme of India. In the first stage, the reactors will use natural uranium fuel (where the uranium-235 component is 0.7 per cent and balance is uranium-238). These reactors called 'thermal reactors' produce power and also covert a small fraction of uranium-238 into plutonium. The plutonium separated by chemical reprocessing and the residual uranium, which is predominantly uranium-238 is to be used in the second stage fast reactors. The fast (neutron) reactors based on plutonium are able to generate plutonium from uranium-238 in excess of that consumed due to typical reactor physics advantages offered by plutonium fuel when fissioned by high energy fast neutrons. These types of reactors are called 'fast breeder reactors', since they deploy high energy neutrons for causing the fission and produce more plutonium than they consume. This physics advantage in a suitably designed fast reactor can thus help to multiply the available fuel and therefore the energy generating capacity at a given time. The process can continue in a repeated manner till the total uranium-238 gets consumed, ensuring long term energy security from a given amount of initial uranium. The process can be carried forward by thorium conversion to uranium-233 in a similar manner as planned for the third stage of Indian nuclear power programme. India has large reserves of thorium on its soil. The thorium based reactors will not breed as much as plutonium does but will be able to sustain the available capacity further for two or three centuries. It may be mentioned here, however, that the technology to handle plutonium and uranium-233 are increasingly complex because of the radioactive nature of these fuels. Thus, the second and third stage of the programme needs additional technology development in this respect and its demonstration at commercial scale.

There are at present 20 reactors in operation in India as part of the first stage of the programme. It may be recalled that India was scientifically/technically cut-off from the world in this technology since 1974, following the first Pokhran experiment.

India has constructed 16 reactors since 1974, entirely based on its own design, engineering, equipment manufacturing, construction, operation and maintenance capabilities available within the country in Department of Atomic Energy, academic institutions, industry, contractors etc. A well demonstrated maturity of handling this technology, safely through integrated 350 years of reactors operation exists.

## Nuclear Safety

In the above introductory para two characteristics of the fission reaction are elaborated *viz.*, capability for high energy output and advantage of simultaneously converting fertile materials into fissile. The fission reaction also produces two smaller nuclei called fission products. While each fission reaction generally produces two fission products, there could be different combinations of these in fission reactions. Overall about 200 fission products of different mass numbers are produced, when uranium-235 atoms fission. Most of these fission products are unstable (radioactive) isotopes of the natural elements. Thus, they tend to become stable overtime by releasing radioactive particles like alpha, beta and gamma rays. The fundamental objective of safety in a nuclear facility is to contain these fission products and the related radiations in a safe manner.

A brief description of the nuclear fuel and associated reactor configuration would help in appreciation of the related safety aspects. Uranium, as it is used in the current nuclear power plants, is used in the chemical form of uranium dioxide, due to its metallurgical and chemical stability at high temperature. The uranium dioxide is used in the form of high density (10.5 grams/cc) cylindrical shaped pellets usually around 1 cm in diameter and length. Large number of such pellets are inserted in a zircaloy tube and sealed welded at either end. Such assembly is referred as fuel rod. Number of fuel rods put together, constitutes a fuel bundle. Such fuel bundles arranged in a specific geometry and configuration and when surrounded by water, constitute a thermal reactor, making sustained fission reactions possible. The fission reaction produces the fission products somewhere inside that solid $UO_2$ pellet. As uranium-235 is a very small fraction of uranium, such fission products in small quantity are embedded in a high density uranium-238 matrix. The fission products particularly those in solid form are thus essentially well bonded in their location. Even after significant number of fission reaction/irradiation, the fuel matrix remains essentially intact to retain the fission products. In terms of nuclear safety, it implies that the fuel pellets and the associated zirconium alloy tube should remain intact at all given times to contain the radioactive fission products, a small fraction of the uranium contained in fuel rods.

The radioactive fission products in the fuel pellet decay to reach a stable configuration by releasing radioactive particles and associated energy at different rates. Some fission products may reach stable nuclei in a short time like seconds and others may do the same over many years. The energetic radioactive particles get absorbed within the fuel as well as other reactor components and the process manifests as heat. Thus, when the plant is operating at power, the heat to be removed from the fuel is that due to the fission energy as well as due to the decay heat of the radioactive fission products. When the fission reaction is terminated during shutdown of a plant by quick addition of neutron absorbing materials like boron carbide or cadmium rods, the heat from radioactive fission products still continues to be present in the fuel. This heat called decay heat, soon following a reactor shutdown is about two per cent of the total heat that was being generated by fuel at power. In simple terms, the nuclear safety is ensured if heat from the fuel rod is adequately removed while it is generating power from fission as well as while the reactor is shutdown. In other words, the nuclear safety involves controlling the reactor heat production to be equivalent to heat removal capability, when the plant is at power, and further remove the produced heat including the residual heat, which is initially two per cent of that at full power and reduces gradually with time. If these heat production and removal actions can be performed reliably, the fuel remains intact resulting into a safe nuclear power.

## Defence In-depth Approach

A multi layered design approach called 'defence in-depth' is used in nuclear power reactors to build reliable means of ensuring safety. The approach in simplified form has the following layers:

- *First Level*: The reactor design including the physics of the configuration is usually so chosen that the configuration is inherently safe and has in built tendency to remain in stable operating domain. The engineering of fuel and the related fuel cooling systems for operation are designed considering all possible loading conditions and have adequate margins and safety. The design loads include those imposed by extreme process conditions as well as those from extreme external postulated events like earthquakes etc. The quality assurance practice envelops all activities of design, manufacturing, construction, operation and maintenance such that the design intent is always met. Thus, in general the attempt is to ensure that in normal operation the systems perform reliably by themselves without failure and without need of any safety action.

- *Second Level*: This level essentially provides features of control in case the reactor power or the cooling provisions indicate a tendency of divergence, particularly towards limiting conditions indicated by design. For this purpose, control rods are provided to control the reactor power and similarly flow and pressure controls are available for maintaining desired cooling. This level also includes shut-off features to quickly terminate the fission reactions in a very reliable manner. The reliability of this capability is maximised generally by deploying two diverse systems for same function using failsafe features. Each shut down system is also having redundant components, thereby reducing heat generation automatically to a minimum level.

- *Third Level*: This level has features to mitigate the consequences of accidents in case they happen. Failures, termed as Design Basis Events/Accidents (DBAs) are considered at the design stage itself in all process systems and their range of operating conditions. Engineering solutions to mitigate effect of each of these DBAs is built into design of the plant and their adequacy demonstrated by well validated methods. Such additionally engineered systems are called safety systems. Thus, in case the cooling of the fuel is not sufficient, independent cooling systems are provided to come into action automatically to maintain the fuel cooling and retain the fission products within the fuel rods. The specific cooling systems (Emergency Core Cooling System) are provided to automatically take over heat removal function in case of failures in main system and have built in redundant power backups, certain amount of initial capability to act without aid of power etc. to ensure reliable operation. These cooling systems may use natural thermosyphon principle, in case of power failure incidences.

- *Fourth Level*: This defence in-depth level is provided to confine the fission products and associated systems, in case these get released from the fuel rod, which might have suffered a damage consequent to complete or partial failure of all the above three levels. This is done by providing a strong containment building, enveloping the reactor and associated systems. Indian plants use a double containment concept, one enveloping other with interspace at vacuum to make it a near zero leak combination. While all actions in above levels will initiate automatically on sensing a distress condition, there are also other complementary features available for operator to manage the incident.

- *Fifth Level*: This defence in-depth is readiness to handle the emergency in case the fission products leak from the confinement into public domain under adverse conditions. These emergency procedures are managed by local government authorities, who obtain inputs from nuclear power plant authorities on extent of radioactive leak, its direction of impact (depending on metrological conditions) for performing actions to assure safety of the people living in the region.

The engineering features provided in level two and three above also deploy features that make them feel safe. Redundant and diverse multiplicity in safety systems including at component level within a system is used to provide high order of reliability. Such safety features are tested regularly to confirm their functioning at desired reliable level. The emergency procedures as in the fifth level are also checked through exercises at fixed intervals by local authorities.

## Conclusion

From the above brief description, it is seen that the multi-level defence in-depth approach also results in physical multi-barrier separation between the public and the fission products. These physical barriers are *viz.*, the fuel matrix, the zircaloy tube over fuel pellets, the high strength boundary of the cooling system, the containment building and the distance between containment building and the public.

With such features in place in all nuclear power plants everywhere, including in India, the radiations received by members of public around the plant is practically negligible when compared to the natural background radiations at any place on the earth. The variations in this natural background from time to time at a place and from place to place are far larger than most conservative estimates of actual radiations received by members of public from a nuclear power plant. The defence in-depth approach provides a strong network of features to make possibility of an accident ending up with release of radioactivity in public domain is an extremely rare event. It may also be mentioned that provisions even under such extremely rare possibility ensure that there is no need for panic, and that sufficient time and means are available to implement the emergency procedures such that there is no resultant impact on health of members of the public.

4

S.K. APTE

# Environmental Impact on Aquatic Ecosystems, Biodiversity, Agriculture and Human Health in the Vicinity of Operating Nuclear Power Plants

Nuclear energy is an efficient, powerful and clean technology that will last the energy needs of our planet, way beyond the fossil fuels. The radioactive products of nuclear fission are contained in the reactor core and do not find their way into the environment. Unlike fossil fuels, the nuclear power production does not involve release of ash, dust or green house gases and do not cause environmental pollution in the conventional sense. The only way the nuclear power plants (NPPs) can possibly impact the environment is either by: (a) marginal increase in the background radiation levels around nuclear power plants, or (b) release of large volumes of heated water, used as coolant in reactor condensers, into the nearby water bodies. The nuclear power plants are subject to, and fully abide by, strict laws and regulations dictated by the appropriate regulatory agencies. In India, the Atomic Energy Regulatory Board (AERB) stipulates the safe radiation limits to radiation workers and to the general public, while the Ministry of Environment and Forests (MoE&F) and the central and state level pollution control boards (PCBs) regulate the thermal discharge into water bodies.

The absolutely low and safe background radiation levels prevalent around all nuclear installations are continuously displayed for the information of the general public and are easy to verify. Yet every now and then, press and media reports highlight the possible adverse impact of radiations and thermal effluents, supposedly released from nuclear power plants, on human health, agriculture, fisheries and environment. A major reason for this, of course, is that unlike other technologies, the disastrous destructive ability of nuclear energy was realised by the world in Hiroshima and Nagasaki, much before its benefits became well known. Any nuclear accidents, caused due to human error or natural disasters, have therefore been treated with great concern and have served as important lessons for the nuclear industry and regulatory agencies to ensure and further enhance environmental and biosafety aspects. Construction of a new nuclear power plant is, therefore, also viewed with great concern and apprehension. Recent unfortunate incident at Fukushima have raised these concerns to unjustifiably alarming levels. One must, however, make a clear distinction between the safety issues related to: (i) normal operation of a nuclear power plant, and (ii) scenario in case of a nuclear accident, such as at Chernobyl, or a natural disaster, such as at Fukushima. This paper addresses the possible impact on aquatic ecosystems, biodiversity, agriculture and human health, in the vicinity of operating nuclear power plants.

## Background Radiation Exposure to Mankind

Ionising radiation is not unique to nuclear industry. From the inception of the earth, the life on this planet, including human populations, has evolved to live with background radiation due to natural radioactivity. Presence of radio-nucleides in the earth's crust or cosmic radiations from the space, all add to the background radiation, we are continually exposed to. The background radiation exposure to humans varies depending on the geographical location and the food and water we consume or the air we breathe (Table 4.1). The national average background dose in India was calculated at 775 $\mu$Gy per year, some time ago and varies widely from state to state and season to season (Nambi *et al.*, 1986; Parthasarathy, 2011). The lowest average dose of 310 $\mu$Gy/year was recorded in Lakshadweep islands, while Andaman islands showed 491 $\mu$Gy/year. Among the states, Maharashtra showed the lowest background dose of 466 $\mu$Gy/year, Andhra and Uttar Pradesh displayed 868 and 985 $\mu$Gy/year respectively, while the highest average dose of 1403 $\mu$Gy/year was recorded in Kerala.

**Table 4.1**

*Background Radiation (µGy/year) and Mankind*

| | |
|---|---|
| *National Average for India per Year* | 775 |
| Lakshadweep | 26 |
| Mumbai | 484 |
| Delhi | 700 |
| Kolkata | 810 |
| Chennai | 790 |
| Hyderabad | 1278 |
| Kerala Monazite Sands | 3800-45000 |
| *Problem of Indoor Radon in UK* | |
| In some areas  5% homes | 23700 |
| 1% homes | 55800 |
| Highest in some houses in Cornwall | 32000 |
| *Uranium Mining in Jaduguda* | |
| Surface of mines | 1000 |
| Tailings Ponds | 1500-2000 |
| *Operating Nuclear Power Plants (at 1.6 km)* | 2-36 |

*Source*:  Based on the information compiled from Parthasarathy (2011) and Rees *et al.* (2011).

Small areas of India, China, Brazil and Iran exhibit very high levels of natural background radiation. Importantly, these areas are inhabited by human populations. The background radiation levels range from 3.5-5.4 µGy/year in Yangziang, China, with a population of 1 lakh people; through 3-35 µGy/year in Guarapari, Brazil with 70000 people residents, 1-45 µGy/year in parts of Kerala and Tamil Nadu, India, with 4 lakh population; to 10-260 µGy/year in Ramsar, Iran, which has only 2000 inhabitants (Hendry *et al.*, 2009). Elsewhere people may be exposed to high natural background radiation due to 222 Radon, an inert gas with a half-life of 3.8 days, formed upon decay of 226 Radium. The gas enters houses through cracks and builds up in poorly ventilated dwellings. For better ventilated dwellings in Kerala, the highest dose measured is 38.4 µGy/year compared to 24-320 µGy/year for poorly ventilated dwellings in some parts of the UK (Parthasarathy, 2011; Rees *et al.*, 2011).

Many of our normal life activities also cause radiation exposure (Table 4.2). For example, one gets as much as 0.4 mSv exposure in 50 air trips of 2 hours duration each or up to 0.1 mSv dose from each chest X-ray. Global average of background radiation exposure to humans stands at 2.4 mSv/year from natural causes. An additional 1.7 mSv/year exposure is estimated on account of medical exposure to humans world wide. Over and above the 2.4 mSv/year background level, the safe radiation exposure limits specified by the AERB are 1 mSv/year to the general public and 30 mSv/year to the radiation workers. In the US, the Environmental Protection Agency has set a limit of 50 mSv/year for radiation worker. Against this backdrop, the public dose from nuclear power plants is extremely small. For the major nuclear power plants (NPPs) in India, the average annual dose over a 5 year period at the 1.6 km radius from the plants ranged from 0.4 to 39.6 $\mu$Sv (Parthasarathy, 2011), which is far below the safe limit of 1000 $\mu$Sv/year specified for common man by the AERB. Exposures above this specified limit are observed only in the high background natural radiation areas or occur to those who venture to live in space stations (170 mSv). The atomic bomb survivors of Hiroshima and Nagasaki were estimated to have received a dose of 200 mSv (Boice, 2012).

**Table 4.2**

*Common Sources of Radiation Exposure*

| Source of exposure | Dose (mSv) |
| --- | --- |
| One chest x-ray | 0.05-0.1 |
| Air travel (50, 2h trips) | 0.4 |
| One year exposure to natural radiation | 2.4 |
| Radiation limits (common man/radiation worker) | 1/20 (50-250) |
| Average Dose at 1.6 km radius of India's NPPs | 0.0004 to 0.0396 |
| Space station per year | 170 |
| A-bomb survivor (mean dose) | 200 |
| Doses used in mutation breeding of plants | 200-400 Gy |
| Doses used in food irradiation | 5-1000 Gy |

## Effects of High Background Radiation Exposure on Human Health

Biological effects of the very low levels of radiations that prevail in the vicinity of NPPs are not discernible. This can be either because no damage is caused at such low levels of radiations or because the living cells can repair, whatever damage is caused by such low levels of radiations. However, the biological effects of radiations are generally viewed in the context of the Linear-No-Threshold (LNT) hypothesis. According to LNT hypothesis, there is no threshold level below which radiations do not cause any damage and can therefore be considered as safe. It does not consider the ability of cells to repair damage caused by very low levels of radiation. Inability to measure damage at low doses could also be due to the possible insensitivity of the methods used. Since, it is also not possible to expose humans to specified low doses of radiations for experimentation, use is made of the human populations living in high level natural radiation areas (HLNRA) in Kerala, to assess the effects. The background radiation dose in Kerala due to monazite sands is much higher (1-45 mGy/year) than what occurs at 1.6 km radius of NPPs (Table 4.1). But such studies provide a basis to calculate possible biological damage at lower radiation doses, by extrapolation.

The small incremental rise in background radiation from NPPs and other nuclear installations or in high background radiation areas (HBRA) has often been wrongly blamed for higher incidence of cancers, down's syndrome and congenital malformations. For example, unbelievably high frequencies of malformations were reported and blamed on radiation from uranium mines and tailings ponds in Jaduguda and NPP at Rawatbhata, by Gadekar and Gadekar in chapter 6 in this volume. But, the investigations at Jaduguda and Rawatbhata were a single time point study each, on a very small population, which limits the statistical validity and significance of the reported findings.

A high frequency of down's syndrome was reported in HBRA of Kerala in the renowned international journal *Nature* (Kochupillai *et al.*, 1976). The study reported 12 down's syndrome cases in 13,000 study population *versus* none in 6,000 normal population.

The statistical validity of this study was subsequently questioned and criticised by Edwards and Harnden (1977) and Sundaram (1977). The final analysis failed to prove that high level natural radioactivity was responsible for the recorded excess of Down's syndrome cases. Similarly, we hear occasional stories of excess cancer incidence in HBRA of Tamil Nadu and Kerala, although there is no scientific evidence of such kind from that region (Rose, 1982). Extensive studies carried out by Bhabha Atomic Research Centre (BARC) scientists, and researchers at the Regional Cancer Centre, Thiruvananthpuram, have similarly found no eveidence of high cancer risk due to external background radiation in monatite-rich areas of Kerala (Nair, 1999).

A detailed long-term analysis of nearly 1,35,000 consecutive newborns in HLNRA populations of Kerala found no correlation between cases of congenital malformations observed at birth or down's syndrome, with radiation levels in this region (Jaikrishan, 2010). A total of 2978 congenital anomalies were detected from which the major malformations constituted about 50 per cent (1496 cases). The overall frequency of congenital malformations in HLNRA was 2.11 per cent as against 2.39 per cent in normal level natural radiation areas (NLNRA). Still births, twins and Down's syndrome cases also showed similar frequencies for both the HLNRA and NLNRA populations. The results from these populations in Kerala are in complete agreement with a nationwide large scale study of malformations and down syndrome incidence (SoMDI) (Table 4.3) carried out in three major cities (Baroda, Mumbai and Delhi) of India (Verma *et al.*, 1998; Modi *et al.*, 1998; Bharucha, 1998). Multiple logistic regression analyses have brought forth a correlation between: (a) still births and maternal age at birth, gravida status and consanguinity, and (b) frequency of malformations and consanguineous marriages (Figure 4.1). However, frequency of malformations innewborns, still births or twinning, showed absolutely no correlation with natural high background radiation. Down's syndrome cases similarly displayed no correlation with radiation levels.

### Figure 4.1

*Multiple Logistic Regression Analyses of Congenital Malformations, Still Births, Down's Syndrome and Numerical and Structural Aberration Found in the Survey of 1,34,178 Newborns in HBRA of Kerala*

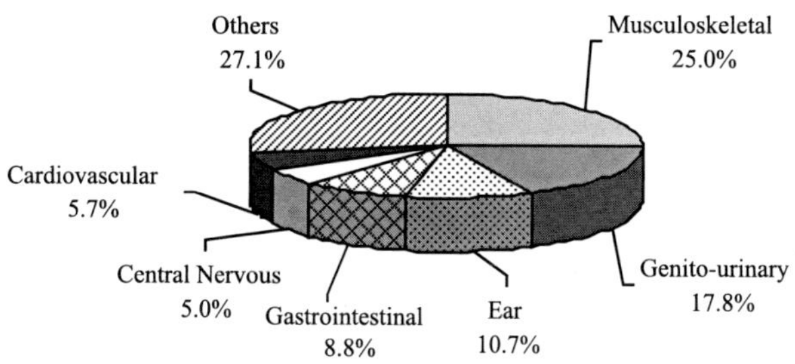


*Logistic Regression Analysis of Congenital Malformation*

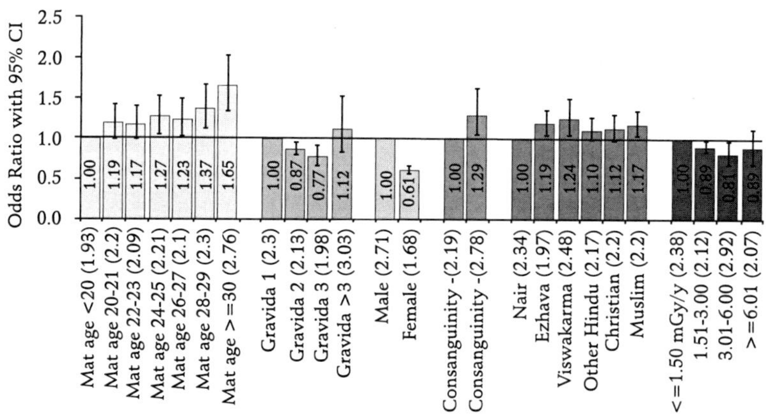


*Down Syndrome and Background Radiation Dose*

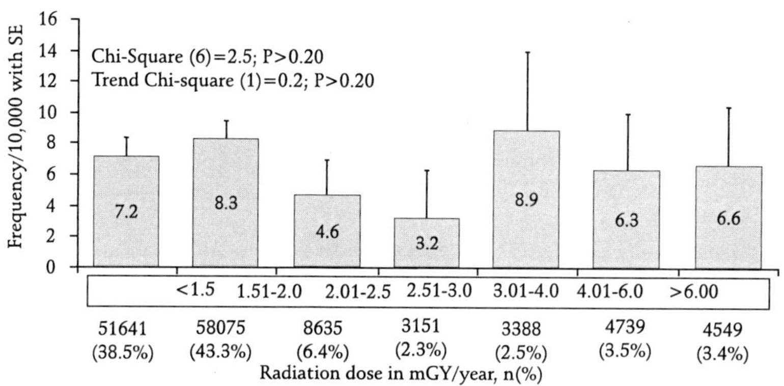

### Logistic Regression Analysis of Stillbirth

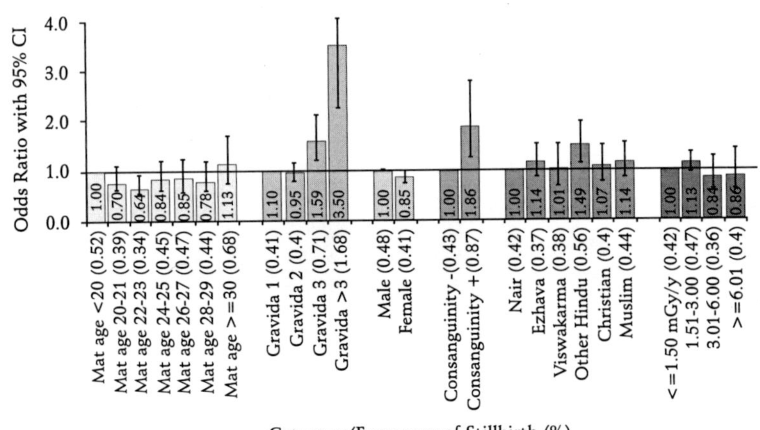



### Frequency of Numerical and Structural Aberrations in Various Dose Groups

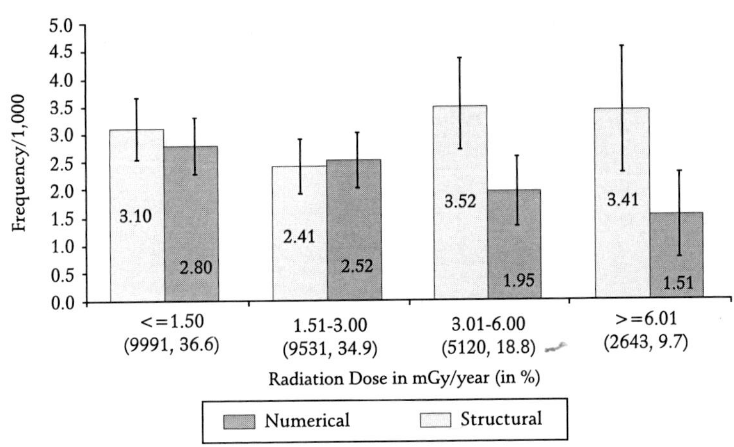



*Note*: A total of 134,178 newborns were monitored for congenital malformations and 2,978 congenital anomalies were detected. The frequency of malformations in HLNRA was 2.11% as compared to 2.39% in NLNRA. No significant difference between HLNRA and NLNRA

*Source*: M. Seshadri & Birajlaxmi Das, RB&HSD, BARC.

**Table 4.3**

*Down Syndrome, Congenital Malformation and Stillbirth Frequency:*
*A Comparison among Delhi, Mumbai, Baroda of SoMDI and HBRA, Kollam*

| Study | n | CA% | DS | SB% |
|---|---|---|---|---|
| SoMDI, Delhi | 23,367 | 1.46 | 1 in 1235 | 3.1 |
| SoMDI, Mumbai | 42,304 | 2.3 | 1 in 1510 | 2.6 |
| SoMDI, Baroda | 31,775 | 2.05 | 1 in 962 | 4.0 |
| Kollam | 134,178 | 2.11 | 1 in 1355 | 0.45 |

*Source*: M. Seshadri, Ex-Head, RB&HSD, BARC.

In recent years, the studies have gone deeper to probe the effects at cytogenetic and molecular levels and have yielded similar results. A total of 27,277 newborns (17,291 from HLNRA and 9,986 from NLNRA) were analysed for karyotype anomalies and about 10,230 newborns (8,493 from HLNRA and 1,737 from NLNRA) were analysed for chromosomal aberrations. The observed frequencies for karyotype anomalies and chromosomal aberrations were similar in the two populations (Cheriyan, 2010). Further probing of binucleated cells with micronuclei of 271 newborns (61 from NLNRA and 210 from HLNRA) did not show significant differences between these two population groups (Das and Karuppasamy, 2009). Similarly, studies on telomere length analysis on 310 adults (233 from HLNRA and 77 from NLNRA) did not show differences in telomere length in HLNRA population as compared to NLNRA (Das *et al.*, 2009). So far, the results obtained from all these biological end points did not show any statistically significant difference between the HLNRA and NLNRA populations, clearly establishing the fact that relatively high doses of radiations (1-45 mGy/year) prevalent in such areas do not cause any adverse effect on human health. What ill-effects can nearly 100 folds lower radiation levels from NPPs can cause to human populations, is anybody's guess. But clearly, there is no basis to associate the negligible radiations from NPPs with any detrimental effects on humans.

## Effect of Thermal Effluents from NPPs on Aquatic Ecosystems

An inherent problem in electricity generation is that all the heat energy cannot be converted into electrical energy and a significant part of it has to be rejected into appropriate heat sinks (Clark, 1969; Parker & Krenkel, 1969). The unavoidable rejection of heat to water bodies needs to be regulated properly so as to avoid undue damage to the aquatic ecosystem of the receiving water body (Hudson & Cravens, 1990, 1991; ITRS-202, 1980). Very strict laws specified by the MoE&F and PCBs, govern such heat dissipation into water bodies. Alarmed by the large scale ocean warming experiences of 1998, regulations regarding thermal effluent discharges were revised and made more stringent all over the world. The new regulations that came into effect in India from June, 1999, stipulated that: (a) the new power plants using water from rivers, lakes and reservoirs shall install cooling waters, (b) new coastal power plants should regulate the thermal discharge in such a way that temperature of receiving water body at the discharge point does not rise by more than 7°C above the ambient temperature, and (c) the existing power plants should ensure that temperature difference at the discharge point should not be more than 10°C above the ambient temperature (Gazette of India Notification G.J.R. 7, dated 22.12.1998). It is pertinent to mention that no data existed in 1998 on the impact of thermal discharges into water bodies in India. In the absence of such data, the new guidelines appeared to be over cautious, but justified. This also brought to attention the need to assess thermal impact on aquatic ecosystems in India.

The Department of Atomic Energy (DAE) took initiative in setting up the first major Thermal Ecological Study (TES) in the country (Apte, 2002; TES Report, 2007), aimed at making a realistic assessment of the biological impact of thermal discharge in the vicinity of power plants. The DAE also offered two of its existing nuclear power plants, one a typical coastal power station at Kalpakkam and the other situated on freshwater Kadra reservoir at Kaiga on the banks of river Kali, as the study sites. A total of 8 laboratories from reputed universities and national research institutes participated in the project, which was co-ordinated and monitored by a committee of experts from National Institute of Ocean Technology (NIOT, DOD), National Institute of Oceanography (NIO, CSIR), Central Electrochemical Research Institute (CECRI, CSIR), Bhabha Atomic Research Centre (BARC,

DAE), Nuclear Power Corporation of India Limited (NPCIL, DAE) and Ministry of Environment and Forests (MoE&F) (Table 4.4).

**Table 4.4**

*DAE-BRNS CRP on Thermal Ecological Studies: Nodal Laboratories, Co-ordinators and Principal Investigators*

---

*Dr S. Narasimhan, WSCL, Kalpakkam & Prof N. Sukumaran, SPKCES, Alwarkurichi*

Baseline studies on thermal ecology of Kudankulam marine environment

*Prof Shahul Hameed, J.M. College, Trichy*

A study on water quality and littoral benthos for impact assessment in the vicinity of MAPS thermal outfall

*Prof V.N. Raja Rao, University of Madras*

Impact of heated water effluent from a coastal power plant on microalgae

*Prof S. Jayachandran, University of Pondicherry*

Microbial ecology in the vicinity of a coastal atomic power plant

*Prof T. Subramaniam, University of Madras*

Distribution and behavioural patterns of the mole cab in the vicinity of MAPS thermal outfall

*Dr P.M. Ravi, ESL, Kaiga*

*Dr T.K. Ghosh, NEERI, Nagpur*

Baseline aquatic ecology of Kadra reservoir, the source of cooling water for Kaiga nuclear power plant

*Dr M.N. Madhyastha, Managalore University*

Baseline studies on benthic ecology and thermal tolerance of some selected species from Kaiga environs

*Dr A.K. Pal, CIFE, Mumbai*

Thermal tolerance of important fish species from Kali river, Karnataka

---

*Source*: Co-ordination and Monitoring Committee: All PIs, NIO, NIOT, CERI, BARC, NPCIL, MoE&F.

The four year duration TES project: (a) measured thermal plume and its distribution around and away from the discharge point at Kalpakkam and Kaiga, (b) evaluated effects of thermal discharge on the water quality and nutrient status of the water body, and (c) assessed biological impact of thermal discharge in the vicinity of power plants. A baseline survey of the physico-chemical parameters and occurrence and distribution of major flora and fauna was also carried out at Kudankulam, the site for two forthcoming nuclear reactors. The study carefully measured the thermal plume distribution, physico-chemical properties of water, and the abundance and distribution of biological forms representing various trophic levels of the ecosystem (Apte, 2002). The data were collected from carefully chosen and global positioning system (GPS-fixed) sampling sites, through regular monthly cruises over a three year period, to cover possible seasonal and spatial variations and to make data comparisons at the site more relevant and meaningful. The observed in situ biological effects of the thermal discharge were further ascertained and validated by additional studies on selected, representative, dominant species in the laboratories, in the fourth year. Details of these results are available in the individual reports of all TES projects submitted to Board of Research in Nuclear Sciences (BRNS) (Ayyappan and Pal, 2005; Ghosh, 2005; Jayachandran, 2005; Madhyastha, 2005; Raja Rao, 2005; Subramoniam, 2005; Sukumaran, 2005).

At both the study sites, the heated effluents from NPPs are carried to the recipient water body either through a man-made discharge canal as in Kaiga, or through a discharge canal created by natural sand bar formation at Kalpakkam (Figure 4.2). Extensive mapping of thermal plume at Kalpakkam (Figure 4.2, inset A) and Kaiga revealed that at both the sites the $\Delta T$ remained within 5-6°C above the ambient temperature, with an occasional $\Delta T$ of 8°C was observed twice at both the sites, during the study period. This was well within the limits specified by MoE&F and thus posed no major harm to the ecosystem. But a $\Delta T$ of 3-5°C did cause limited biological damage, which remained contained in a small mixing zone at the point of confluence of thermal discharge with the water body (TES Report, 2007). The study carefully mapped the physical boundaries of such mixing zone at both the sites. In general, the thermal plume was buoyant and remained attached to 1-3m depth near the surface. At Kalpakkam, where extensive turbulent mixing of discharged water with seawater takes place due to tidal activity, a $\Delta T$ of 3-5°C is observed in a 500m x 200m area beyond the mixing point, while a $\Delta T$ of 1-3°C prevails in a 500m x 300m area next to the first zone (Figure 4.2, inset B). The position of this mixing zone shifts seasonally, up to a distance of 1.2 kms, due to formation of a natural sand bar at Kalpakkam, but at any given time its dimensions remain the same as shown in the figure. In certain months, the first zone extends up to a linear distance of 1 km along the shore with a width of <50m (TES Report, 2007). At Kaiga, due to relatively stagnant nature of the reservoir, the heat dissipation is relatively slow. Since the annual variation in surface water temperatures is very large at Kaiga (23-31°C), the mixing zone dimensions are very different in summer and in winter (Figure 4.3). A ΔT of 5-6°C is observed in a 350m x 125m zone beyond the end of the discharge canal (EDC), which decreased to a ΔT of 3°C in a 450m x 250m zone. Beyond the 500m x 300m zone, the ΔT was less than 2°C. In summer, the highest ΔT observed was only 3°C and was restricted to a 250m x 50m zone in front of the EDC (Figure 4.3). At Kaiga, these were the dimensions of the mixing zone in the two seasons (TES Report, 2007).

**Figure 4.2**

*Temperature Variation Across the MAPS Outfall, Discharge Canal and Mixing Zone*

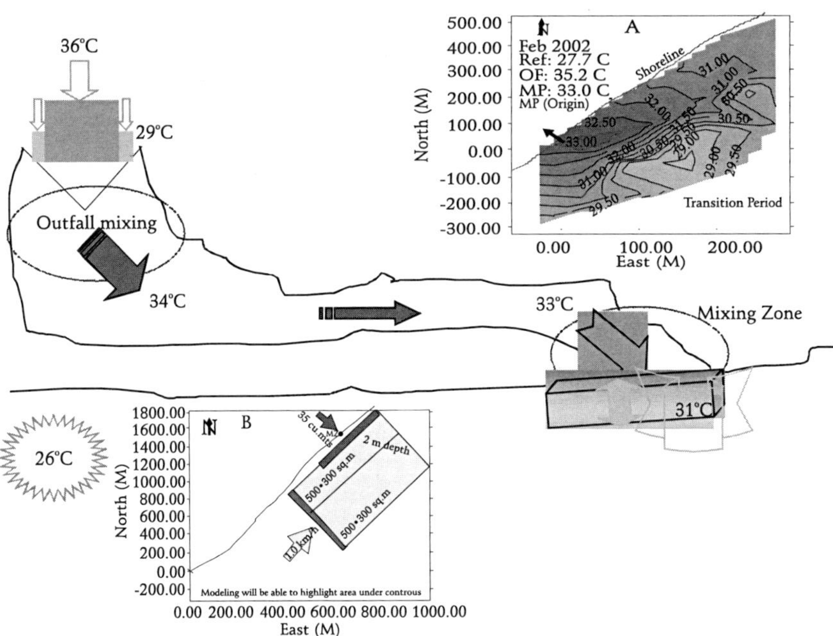


Note: Thermal plume measurements were made every fortnight over a period of 3 years. *Inset A*: Representative distribution of isotherm beyond the mixing zone in Kalpakkam in February, 2002. The intake, outfall and mixing point temperatures are indicated. *Inset B*: Physical mapping of mixing zone in Kalpakkam waters based on 4 year long thermal plume measurements.

*Source:* S. Narasimhan, Ex-Head, WSCD, BARC.

**Figure 4.3**

*Temperature Variation Beyond the End of Discharge Canal and the Physical Limits of the Mixing Zone at Kaiga*

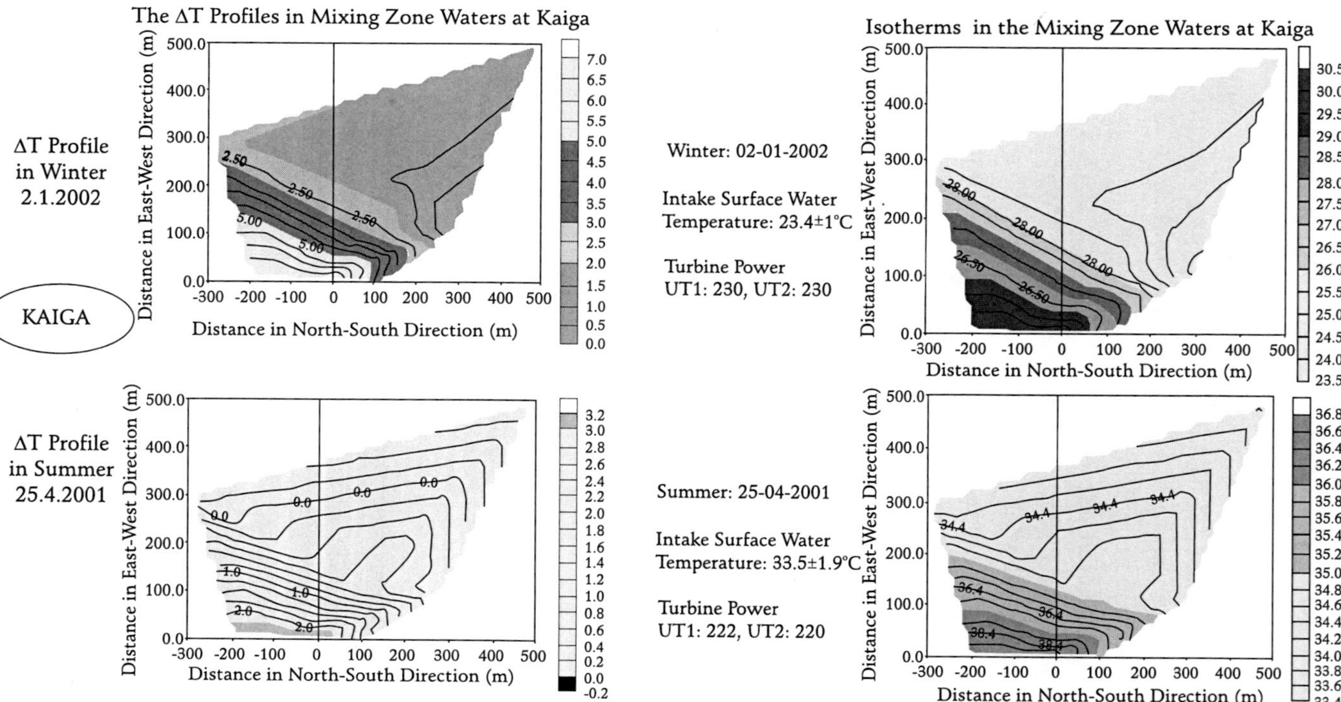



*Note:* The ΔT profiles and isotherm distribution are shown for a given typical day in winter and summer months. The reference intake water temperature and turbine power on the sampling day are indicated.

*Source:* P.M. Ravi, Ex-Head, ESL, Kaiga.

At both the sites, biota were examined at all trophic levels and included primary producers (phytoplankton, cyanobacteria and algae), consumers (zooplankton, insects, mollusks and fishes) and decomposers (bacteria and fungi) (Ayyappan and Pal, 2005; Ghosh, 2005; Jayachandran, 2005; Madhyastha, 2005; Raja Rao, 2005; Subramoniam, 2005; Sukumaran, 2005). An adverse impact on the biota was seen at the discharge point or Outfall, both in Kalpakkam and Kaiga sites. At Kalpakkam, the most affected were the phytoplankton, microalgae and bacteria at the Outfall (Jayachandran, 2005; Raja Rao, 2005), and the benthic forms such as bivalves and mole crabs (Hameed, 2005; Subramoniam, 2005) up to a limited distance along the shoreline, on either side of the mixing point. Barring the benthos, abundance and distribution of most life-forms returned to near normal level beyond the mixing zone at Kalpakkam. Comparison of limited fishing carried out in Kalpakkam waters (Hameed, 2005), with neighbouring fishing sites at Mahabalipuram or Sadras revealed no significant adverse effects on fish abundance or diversity at Kalpakkam (TES Report, 2007).

At Kaiga, the most affected were the phytoplankton and zooplankton (their abundance somewhat decreased up to a linear distance of 500m from EDC) and the benthos and bacteria in the mixing zone (Ghosh, 2005; Madhyastha, 2005). At Kaiga the seasonal variation in surface water temperature is large (winter 23°C and summer 31°C). Beyond the mixing zone at Kaiga, although the $\Delta T$ values were less than 3°C, the absolute temperature depended on the season. And normalcy of life was restored only when biologically conducive temperature values were established (TES Report, 2007). This is best illustrated by the primary productivity (Figure 4.4), which was near normal even at the EDC in winter (30°C) but was restored to normal only beyond 500m from EDC in summer (32°C). Thus apart from $\Delta T$, the absolute temperature is even more important in determining the diversity and abundance of life forms and it needs to be taken into account by the regulatory agencies. At Kaiga also, the life was restored to nearnormal beyond the mixing zone. No adverse effect on fish population or fish catch was observed at Kaiga (Ayappan and Pal, 2005), which is located adjacent to a major fishing resource, the Kadra reservoir.

**Figure 4.4**

*Primary Productivity (By $^{14}$C Method) at Selected Sampling Points Beyond the End of Discharge Canal in the Kadra Reservoir*

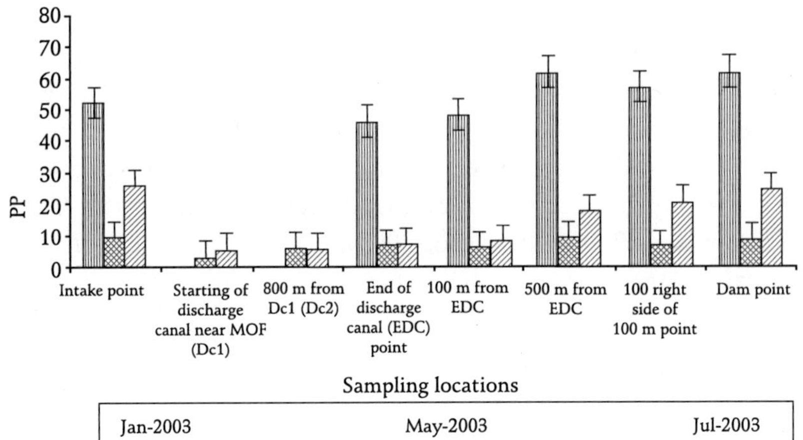


*Note*: Estimation of primary productivity at Kaiga in winter (January), summer (May) and monsoon (July) seasons in 2003. Note the restoration of primary productivity right at the end of discharge canal in winter and beyond a distance of 500m in summer months.

*Source*: T.K. Ghosh, NEERI, Nagpur.

All the physico-chemical, biological and thermal data clearly show that the thermal discharge from the power plants should be considered as merging with the receiving water body in a 'mixing zone' created at the point of confluence of discharge. The concept of mixing zone has been described earlier also (Gajendragadkar and Shirvaikar, 1998). The physical dimensions of this mixing zone in terms of permissible $\Delta T$ and absolute temperature have been mapped at both the sites (Figures 4.2 and 4.3) in TES (TES Report, 2007). For a coastal site like Kalpakkam with less annual variation in ambient temperature and extensive mixing at the EDC, the zone dimensions do not vary much. But at Kaiga, the extent of mixing zone clearly varies with season, being smaller in winter and much larger in summer. The results suggest that the mixing zone dimensions can be site-specific and seasonally determined. One can thus visualise a 'dynamic mixing zone' that is determined by the nature of water body, geographical location, seasons, power plant design and nature of prevalent biota. Indeed, the practice of a multi-factorially determined mixing zone is in vogue in the Western countries (Essig, 1998)) and can be practiced in India as well.

An ecologically encouraging finding of this study was the fact that the observed ill-effects of thermal discharge are not very extensive but are restricted to a very small area and appear to be reversible. In particular, the seasonal shifting of the sand bar and mixing zone at Kalpakkam and the large seasonal variation in ambient temperature at Kaiga allowed restoration of parameters (especially temperature) to biologically more conducive levels. Under these situations, the impacted sites were recolonised and reinhabited by the forms that were adversely impacted otherwise (Figure 4.5). This is best illustrated by the recolonisation of

**Figure 4.5**

*Recolonisation of the Shifting Mixing Zone by the Bivalve*
*(Donax) at Kalpakkam*

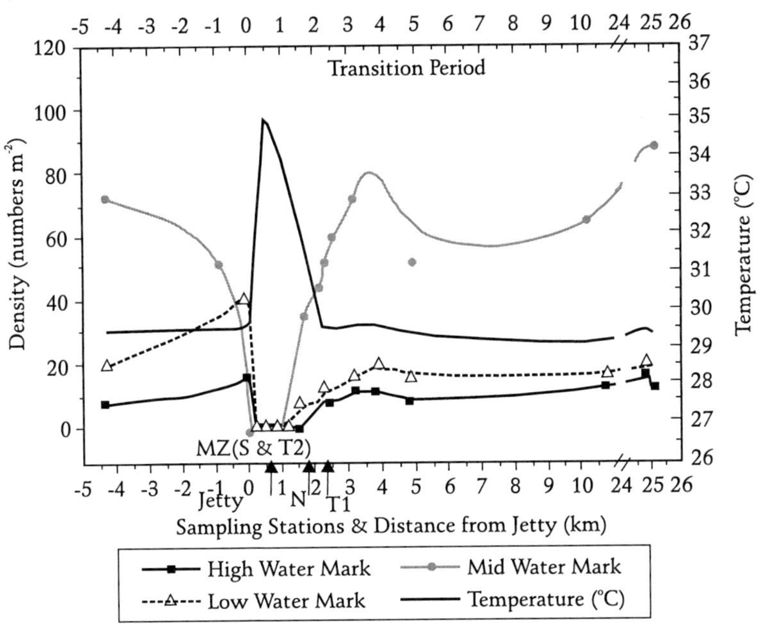


*Note:* *Donax sp* density at three water marks at intertidal zone in relation to sea water temperature during February 2003.

*Source:* Shahul Hameed, J.M.M. College, Thiruchirapally.

earlier mixing point by bivalves and molecrabs (Hameed, 2005; Subramoniam, 2005) at Kalpakkam, when the mixing point shifs seasonally. Thus, the biological impact of thermal effluents clearly seem to be restricted in space and reversible in time.

The salient findings of the TES are in broad agreement with similar studies conducted in other parts of the world (Hawkes, 1969; Langford, 1990: Santhanam *et al.*, 1980; Suresh *et al.*, 1993). A film, entitled "And quiet merges the warmth...", based on the scope, work plan and salient findings of the Theram Ecological Studies was prepared by the Central Institute of Fisheries Education (CIFE), Mumbai, as part of this project. The film and the TES project report were released by the DAE-BRNS in March, 2011.

## Impact of Nuclear Power Plant Operation on Biodiversity and Agriculture

In recent years, there have been press reports stating that construction of nuclear power plants at Kudankulam or Jaitapur will destroy biodiversity. Nothing can be more far from truth than this misinformation, and those who have visited our nuclear power plants are a witness. Who can deny that the Kaiga power station located amidst a thick forest in Western Ghats is in perfect harmony with the pristine beauty of its surroundings? A pleasant recent surprise is the less known fact that the avian biodiversity at the Narora Atomic Power Station has substantially increased in last few years, while in the same period the diversity of visiting birds at the nearby Bharatpur bird sanctuary has declined. The aforesaid instances are excellent examples of biodiversity sustenance efforts at all the NPPs. Industrialisation has often led to adverse effects on biodiversity and agriculture in the past, not only in terms of loss of arable land but also through the consequent physico-chemical pollution. This is also true of electricity generation by thermal power stations which generate a lot of ash or green house gases. But nuclear power plants are a notable exception.

As mentioned earlier, nuclear power is a clean, no-ash, no-GHG (green house gas) technology and does not pollute the environment in any way, except for the heated coolant water discharge in the nearby water bodies. This can potentially impact aquatic ecosystems, especially the mangroves, fish and corals of great commercial interest to mankind as ecological niches of useful bio-resources and for bioprospecting, as food and feed, and as sites of tourist attraction. Site selection for NPPs always takes into account and avoids proximity to large mangrove communities or coral reeves. There are no major coral reeves or mangrove forests in the close vicinity of any existing or forthcoming nuclear installation in India. At the BARC coastline in Mumbai, mangroves and nuclear reactors have been in perfect equilibrium and peaceful co-existence for over 50 years now (Figure 4.6). As detailed in the above section, limited adverse impact in response to thermal effluents discharged was seen at Kalpakkam and Kaiga on planktons and benthos, as a consequence of their extremely poor motility. But this is restricted to a negligibly small area and has no consequence for the recipient water body at large, as evidenced by the fact that beyond the

**Figure 4.6**

*Peaceful Coexistence: Mangroves and Reactors for 60 Years*

*Note*: Mangrove belt along the BARC coastline. The mangroves have co-existed with the two research reactors Dhruva and Cirus at BARC, Mumbai for last several decade.

*Source*: K. Bhanumurthy, Head-SIRD, BARC, Mumbai.

mixing zone, the life is near normal (TES Report, 2007). Even in the mixing zone, the impact is not permanent but is reversible as demonstrated by the recolonisation of such sites upon restoration of conducive environment in different seasons at Kaiga and by the natural shifting of mixing zone at Kalpakkam.

Fishes are highly motile and have an excellent nervous system. Laboratory studies on thermal avoidance behavior of native fish of Kalpakkam, Kaiga and Kudankulam waters have demonstrated that they can perceive even small thermal changes well and respond to them by rapidly moving away from adverse situations (Ayyappan and Pal, 2005; Sukumaran, 2005). They also exhibit a wide range of thermal tolerance. At coastal sites, such as Kalpakkam and Kudankulam, the annual variation of surface water temperature is very small (26-30°C). Even at Kaiga, where such variation is large (23-33°C), the minimum and maximum critical temperatures for fishes resident in Kadra reservoir were found to be 11-18°C (CTmin) and 38-42°C (CTmax) (Table 4.5) (Ayyappan and Pal, 2005). Such temperatures are never reached in the reservoir. Therefore, there is no scientific basis for the misinformation that fisheries are or would be seriously affected as a result of nuclear power plant operation. This has indeed been corroborated by the fishing studies carried out at

**Table 4.5**

*Critical Temperature (CT Max and CT Min) of Fish Species of Kadra Reservoir (Ambient Temperature during the year 23-33ºC)*

| Species of Fish | CTMax(ºC) | CTMin (ºC) |
| --- | --- | --- |
| Gonoproktopterus curmuca | 41.10±0.08 | 20.54±0.05 |
| Danio aequipinnatus | 38.00±0.22 | 15.00±0.44 |
| Etroplus suratensis | 42.60±0.05 | 17.00±0.05 |
| Horabragrus brochysoma | 41.01±0.08 | 15.14±0.05 |
| Labeo calbasu | 40.10±0.09 | 17.82±0.03 |
| Mastocembelus armatus | 39.00±0.08 | 12.14±0.05 |
| Ompok malabaricus | 40.03±0.14 | 13.08±0.13 |
| Parluciosoma doniconius | 39.00±0.09 | 11.06±0.09 |
| Puntius filamentosus | 38.53±0.12 | 14.00±0.05 |

*Source*:  A.K. Pal, CIEF, Versova, Mumbai.

Kalpakkam and Kaiga sites, which showed no change in fish diversity or abundance (Hameed, 2005; Ayyappan and Pal, 2005). This should allay all unfounded fears in this regard. In fact small temperature upshift has been found to enhance fish hatching. To make use of this observation, a fish hatchery has been established at Kaiga and makes use of the warm water from the discharge canal to enhance fish aquaculture (Figure 4.7). The hatchery has been in operation at Kaiga for the last 3-4 years with beneficial consequences.

**Figure 4.7**

*Fish Hatchery at Kaiga*

*Note*: Fish hatchery established at Kaiga. The hatchery makes use of warm water from the discharge canal to enhance hatching of fingerlings.

*Source*: A.K. Pal, CIEF, Versova, Mumbai.

The marginal increment in the background radiation levels around NPPs in India (2-40 $\mu$Sv) is too low to cause any ill effect even on human populations, leave aside on the microbial and plant life. Bacteria easily survive nearly 1000-fold higher radiation dose (>1 kGy) than humans (6 Gy) can, and organisms such as Deinococcus radiodurans can withstand even 6-15 kGy of $^{60}$Co $\gamma$-rays (Basu and Apte, 2012). Plants, although they do not match bacteria in radioresistance, display sufficiently higher radiation tolerance than humans and other animals. The radiation-induced mutagenesis of plants, aimed at improving crop varieties, generally employs radiation doses of 200-400 Gy on plant seed material, to induce one or two useful scorable mutations. To date, BARC has produced more than 40 improved crop varieties in several different crops using this technique (Table 4.6). Obviously, therefore, microbial and plant diversity cannot be affected by negligible radiation levels that prevail in the vicinity of nuclear power plants.

**Table 4.6**

*Improved Crop Varieties Developed at BARC*

| | |
|---|---|
| Groundnut | 14 |
| Mustard | 3 |
| Soybean | 2 |
| Blackgram | 4 |
| Greengram | 7 |
| Pigeonpea | 3 |
| Cowpea | 1 |
| Sunflower | 1 |
| Rice | 1 |
| Jute | 1 |

*Notes*: i) Using radiation induced mutations, 39 crop varieties have been developed and released for commercial cultivation in different agro-climatic zones in the country. Some of the varieties are very popular and grown extensively The improved characters are higher yield, earliness, large seed size, resistance to biotic and abiotic stresses.

ii) Doses Used: 200-400 Gy of 60 Co g-rays.

iii) Radiation-induced mutagenesis has been used at BARC to produce improved varieties of several crops. About 40 of these have been released by the ICAR as superior varieties with higher yield and abiotic and biotic stress tolerance.

*Source*: S.F. D'Souza, Head-NA&BTD, BARC, Mumbai.

A positive feature of nuclear power plant construction in India is the practice of first developing a green belt at the site. Right from the first atomic power station at Tarapur to the new, about to be commissioned nuclear power plants at Kudankulam, this practice has been in vogue. At all the NPPs, lakhs of trees have been planted to develop green belt through the efforts of DAE scientists and at times by non-DAE agencies, such as M.S. Swaminathan Research Foundation, Chennai. Anyone who has visited these sites before and after the construction of NPPs, can easily vouch for the fact that the erstwhile barren lands at these sites have been turned in to beautiful green locations because of such efforts. This should also eliminate the apprehension that plant life or agriculture may be adversely affected in the vicinity of nuclear power plants. As a matter of fact, within the 1.6km exclusion zone of many of our NPPs, agriculture and horticulture is practiced rather intensely. Often it involves use of improved crop varieties of groundnut, mustard, rice and various pulses, developed by BARC (Figure 4.8). A whole variety of trees like banana, guava, sapota, mango (including the elite Alphonso, Langda and Dassehari varieties), and coconut are abundantly found in the NPP campuses from Narora, Tarapur to Kaiga and Kalpakkam (Figure 4.9). One misinformation that was doing rounds in Konkan region of Maharashtra last year was that opening of a

**Figure 4.8**

*Agriculture in Practice at TAPS, Tarapur*

*Note:* Successful cultivation of rice, mustard, groundnut and various pulses is regularly practiced at the site by farmers, through a cooperative scheme.

*Source:* S.F. D'Souza, Head-NA&BTD, BARC, Mumbai.

**Figure 4.9**

*Fruit Trees Bearing Coconut, Bananas, Chikku, and Guavas at TAPS, Tarapur*

*Source*: S.F. D'Souza, Head-NA&BTD, BARC, Mumbai.

**Figure 4.10**

*Irradiation Facility at Lasalgaon (Nasik) and Export of Mangoes to the US*

2007: 157 Tons
2008: 275 Tons
2009: 130 Tons
2010: 100 Tons
2011: 100 Tons

Doses used : 250–1000Gy γ-rays

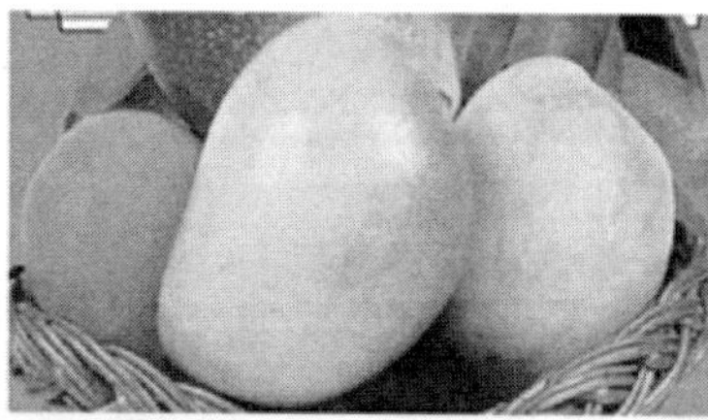

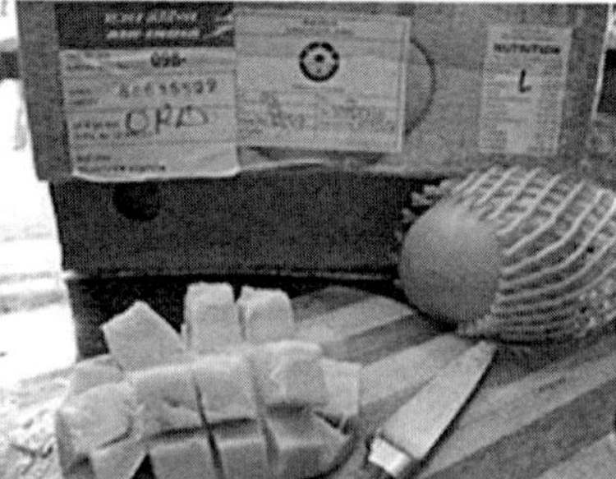

*Note*: The BARC irradiation facility at Lasalgaon, Nasik is used for disinfestations of Alphonso mangoes for their export to USA.

*Source*: A.K. Sharma, Head-FTD, BARC, Mumbai.

NPP at Jaitapur will result in poor quality of the famous Alphonso mangoes. It should suffice to say that in the last 3-4 years, export of Alphonso mangoes to USA has started, only on the condition that the fruits are irradiated in the KUSHAK irradiation facility of BARC at Lasalgaon near Nasik (Figure 4.10). The dose of $\gamma$-rays ($>$1kGy) used for fruit disinfestation is at least a million times higher than the background radiation that prevails in the vicinity of any NPP.

## Conclusion

This paper has attempted to detail information on human health, biodiversity and agriculture in the vicinity of operating nuclear power plants in India, especially on aspects about which some apprehensions have been expressed in certain sections of the press or in media reports or in some limited studies in recent years. If observations contrary to what is presented here should come to notice in future, it would be best to address them through joint studies carried out by DAE scientists and non-DAE independent experts. DAE would be happy to support such relevant studies.

## References

Apte, S.K. (2002). *Thermal Ecology* (B. Venkataramani and N. Sukumaran, eds.) B.R.N.S., DAE, Mumbai, pp 57-60

Ayyappan, S. and A.K. Pal (2005). *Thermal Tolerance of Important Fish Species from Kali River, Karnataka*. Report submitted to BRNS, DAE, as part of CRP on TES.

Basu, B. and S.K. Apte (2012). *Mol. Cell Proeomics.*

Bharucha, B.A. (1998). *Ind. J. Human Genet.* 4: 88-92.

Boice, J.D. Jr. (2012). "Radiation Epidemiology: A Perspective on Fukushima", *J. Radiol. Prot.* 32: N33–N40.

Cheriyan, V.D. *et al.* (1999). "Genetic Monitoring of the Human Population from High-level Natural Radiation Areas of Kerala on the Southwest Coast of India" and "Incidence of Numerical and Structural Chromosomal Aberrations in the Lymphocytes of Newborns", *Radiat Res.* 152 (6 Suppl): S154-8.

Cheriyan, V.D. (2010). "Cytogenetic Studies on Newborns from High Level Natural Radiation Areas of Kerala in the Southwest Coast of India. Presented in *7th International Confrenece on High Level of Natural Radiation and Radon Areas.* Held at The Park Hotel, Navi Mumbai, India, from November 24-26, 2010, OL32, pp. 80-82

Clark, J.R. (1969). *Sci. Amer.* 220: 1-10.

Das, B. and C.V. Karuppasamy (2009). *Int J Radiat Biol.* 85: 272-80.

Das, B., D. Saini and M. Seshadri (2009). "Telomere Length in Human Adults and High Level Natural Background Radiation", *PLoS One.* 4 : e8440

Edwards, J.H. and D.G. Harnden (1977). *Nature* 267: 728-729.

Essig, D.A. (1998). The Dilemma of Applying Uniform Temperature Criteria in a Diverse Environment : An Issue Analysis". Report from Idaho Division of Environmental Quality, Idaho, USA.

Gajendragadkar, S. and V.V. Shirvaikar (1998). "Thermal Pollution Regulations: Need for Sensible Approach in Conservation & Management of Aquatic Resources", (A. Gautam, ed.). New Delhi: Daya Publishing House. pp 162-172.

Ghosh, T.K. (2005). *Baseline Aquatic Ecology of Kadra Reservoir, the Source of Cooling Water for Kaiga Nuclear Power Plant.* Report submitted to BRNS, DAE, as part of CRP on TES.

Hameed, S. (2005). *A Study on Water Quality and Littoral Benthos for Impact Assessment in the Vicinity of MAPS Thermal Outfall.* Report submitted to BRNS, DAE, as part of CRP on TES.

Hawkes, H.A. (1969). "Engineering Aspects of Thermal Pollution" (F.L. Parker and P.A. Krenkel, eds.). Vanderbilt University Press, USA, pp. 15-57.

Hendry, J.H. (2009). *J. Radiol. Prot.* 29: A29–A42.

Hudson, J. and J.B. Cravens (1990). *Research J. Water Pollut. Contr. Fed* 62: 558-568.

———. (1991). *Research J. Water Pollut. Contr. Fed.* 63: 593-607.

ITRS (1980). *Environmental Effects of Cooling Water Systems : Report of a Co-ordinated Research Programme on Physical and Biological Effects on the Environment of the Cooling Systems and Thermal Discharges from Nuclear Power Stations.* ITRS-202 : IAEA Technical Report Series No. 202. International Atomic Energy Agency,Vienna

Jaikrishan, G. *et al.* (1999). "Genetic Monitoring of the Human Population from High-level Natural Radiation Areas of Kerala on the Southwest Coast of India and Prevalence of Congenital Malformations in Newborns", *Radiat Res.* 152 (6 Suppl): S149-53.

Jaikrishan, G. *et al.* (2010). "A Prospective Study on Congenital Malformations in the High Background Radiation Areas of Kerala". Presented in *7th International Conference on High Level of Natural Radiation and Radon Areas.* Held at The Park Hotel, Navi Mumbai, India, from November 24-26, pp. 37-38.

Jayachandran, S. (2005). *Microbial Ecology in the Vicinity of a Coastal Atomic Power Plant.* Report submitted to BRNS, DAE, as part of CRP on TES.

Langford, T.E. (1990). *Ecological Effects of Thermal Discharges.* Elsevier Applied Sciences, London, pp. 28-103.

Madhyastha, M.N. (2005). *Baseline Studies on Benthic Ecology and Thermal Tolerance of Some Selected Species from Kaiga Environs.* Report submitted to BRNS, DAE, as part of CRP on TES.

Modi, U.J. *et al.* (1998). *Ind. J. Human Genet.*4: 93-98.

Nair, M.K. *et al.* (1999). *Radiat Res 152* (6 suppl): S145–S148.

Nambi, K.S.V. *et al.* (1986) *Natural Background Radiation and Population Dose Distribution in India.* Bhabha Atomic Research Centre, Mumbai.

Parker, F.L. and P.A. Krenkel (1969). *Physical and Engineering Aspects of Thermal Pollution,* Vanderbilt University Press, USA.

Parthasarathy, K.S. (2011). *INS Newsletter* 8: 42-51.

Pillai, N.K. *et al.* (1976). *Nature 262.* pp 60-61.

Rao, V.N. Raja (2005). *Impact of Heated Water Effluent from a Coastal Power Plant on Microalgae.* Report submitted to BRNS, DAE, as part of CRP on TES.

Rees, D.M., E.J. Bradley, B.M.R.G. Green (2011). *HPA-CRCE-015: Radon in Homes in England and Wales: 2010 Data Review.* Chilton, UK: HPA.

Rose, K.S.B. (1982). *Nuclear Energy* 21: 399-408.

Santhanam, R., N. Sukumaran and M.N. Kutty (1980). *Indian J. Marine Sci.* 9: 296-297.

Suresh, K. *et al.* (1993). *Hydrobiol* 268: 109-114.

Subramoniam, T. (2005). *Distribution and Behavioral Patterns of the Mole Crab in the Vicinity of MAPS Thermal Outfall.* Report submitted to BRNS, DAE, as part of CRP on TES

Sukumaran, N. (2005). *Baseline Studies on Thermal Ecology of Kudankulam Marine Environment.* Report submitted to BRNS, DAE, as part of CRP on TES.

Sundaram, K. (1977). *Nature* 2: 728.

TES Report (2007). *Thermal Ecological Studies: A DAE-BRNS Co-ordinated Research Project Report.* BRNS, DAE, Mumbai pp 127

Venkataramani, B. and N. Sukumaran (eds.) (2002). *Thermal Ecology.* BRNS-DAE, Mumbai

Verma, I.C. *et al. (1998). Ind. J. Human Genet.* 4: 84–87.

*Acknowledgements:*

The valuable inputs from my colleagues M. Seshadri, Birajlaxmi Das, Anu Ghosh, S.F. D'Souza and A.K. Sharma from the Biomedical Group; Drs D.N. Sharma and R.M. Tripathi from the Health and Safety Group; and Dr K. Bhanumurthy, Head-SIRD of Bhabha Atomic Research Centre, Mumbai, in the preparation of this manuscript, are gratefully acknowledged. Sincere thanks are also due to all the investigators and experts who contributed to the DAE-BRNS CRP on Thermal Ecological Studies (Table 4.4), and to INSA for inviting me to participate in the symposium on Challenges in Nuclear Safety.

5

P.C. KESAVAN

# Low Dose Radiation Health Hazards

## *Radiobiological Mechanisms versus Theoretical Predictions based on Linear, Non-threshold Model*

## Introduction

There is absolutely no doubt, whatsoever, that precautionary principles and safety measures ought to be of paramount importance in any human endeavour/enterprise, especially when there are adverse health implications not only to the persons directly affected, but also to their succeeding generations. Yet, it is also equally important to avoid unfounded apprehensions and justifying them by resorting to methods not based on hardcore science and technology, but by untenable extrapolations and unreasonable appeals. This indeed is the current scenario with reference to radiation risk assessment and determination of radiation protection standards following exposures to low doses of ionising radiation. Soon after the twin discoveries of X-rays and radioactivity, made only a few months apart, a few years before the close of the 19th century, cancer incidence among the radiation researchers became evident. That exposure to ionising radiation could result in cancers was known as early as 1902. At that time, it was not known that DNA is the genetic material, and that cancer possibly results from deleterious changes in some of the genes. Consequently, when in 1925, protective limits (doses of exposure) were suggested for the safety of nuclear workers, the concept of 'tolerance dose' formed the basis. The tolerance doses were determined on the basis of appearance of 'deterministic effects', such as epilation, reddening of skin, bone-marrow depression etc., which require absorption of specific moderate to high doses of radiation by the tissues concerned. The tolerance doses were prescribed within the limits of doses, which cause specific deterministic effects.

The epoch-making discovery in 1927 by H.J. Muller that X-rays induce sex-linked recessive lethal mutations (now known as deletions of DNA segments) in fruit flies (*Drosophila melagnogaster*), for the first time revealed ionising radiation as a mutagen. It was not, however, established then that mutagens, in general, are also carcinogens. The noteworthy point is that even after the new knowledge gained in 1927, the radiation protection standards were continued on the basis of tissue tolerance dose limits. It was only after the atomic bombs detonations over Hiroshima and Nagasaki in August 1945, that H.J. Muller's findings about two decades earlier, started to receive increasing attention. The deep concern centred around the possible heritable genetic damage transmitted from the atomic bomb survivors to their children. Further, the epidemiological studies during the 1950s on the occurrence of leukaemia in children exposed to radiation from the atomic bombs, led to a notion that for the genetic effects, there was probably no threshold dose, and there possibly exists a linear relationship between the dose of exposure and leukaemia incidence. This study covered children at different ages (some were in their mother's wombs), when the atomic bombs exploded and released radiation. Since, there is a lag period between absorption of radiant energy and appearance of cancer, it was noted that the radiation–induced genetic effects are the 'late' biological effects ('late effects') unlike the deterministic effects that normally appear within a few hours to days. The heritable genetic effects would take much longer periods, say even succeeding generations, to manifest. The genetic effects in the somatic (cancer) and germinal (heritable) cells were separated from the 'deterministic effects' and classified as 'stochastic effects' (the term stochastic literally means 'random'). These are illustrated in Figure 5.1.

## Figure 5.1

*Stochastic and Deterministic Effects*

*(a) Stochastic Effects: The Probability of Occurrence of Genetic Effects (Cancer) Increases with Dose*

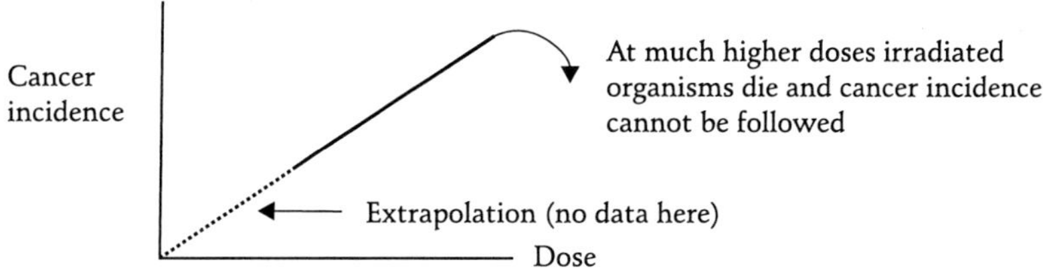


*(b) Deterministic Effects: There is No Appearance of Cancer until Certain Dose of Radiation is Absorbed*

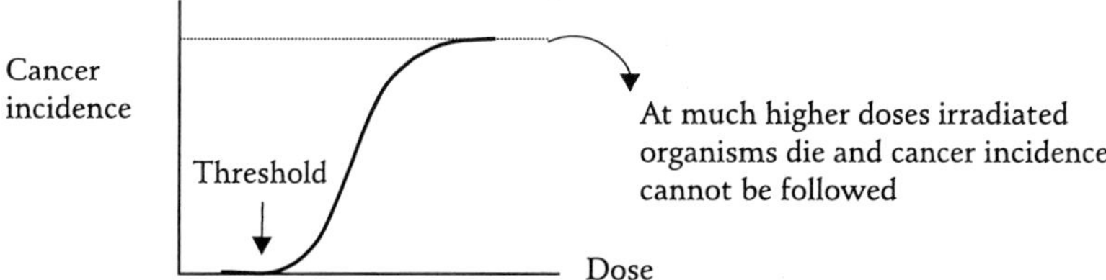


In both the models, death of the individuals exposed to high doses precludes the appearance of cancer.

What the stochastic effects have come to mean today is that there is no threshold level of radiation exposure, below which it could be stated with certainty that cancer or heritable changes will not occur. It is also necessary to point out that the stochastic effects (i.e. cancer and heritable genetic changes) at very low doses are not really experimentally demonstrable and therefore, are estimated (predicted) by backward extrapolation from the effects observed at high doses. About five decades ago, Glass and Ritterhoff (1961) published a paper in Science on the mutagenic effect of a 5R (0.05 Gy) dose of X-rays in *Drosophila melanogaster*. They examined more than a million flies, about half of which were controls that received no radiation, the other half being progeny of flies receiving 5R (R = Roentgen Units) (0.05Gy) of X-rays. The authors first made a backward extrapolation from established dosage curves at 1000R(10 Gy) and 2000R (20 Gy), assuming a linear proportionality of mutations to dose and estimated the mutations experimentally induced by 5R (0.05 Gy). Their estimation suggested an increase in the mutation rate of 0.005 per cent, in comparison with a spontaneous rate 0.04 per cent . In the actual experimental count, control group of more than one half million flies had 283 mutations; and the irradiated group of similar number of flies had 323 mutants. The overall control rate was 0.048 per cent, while their irradiated group showed 0.055 per cent, or an increase over the control of 0.007 per cent, which was not statistically significant. Their conclusion, however, was that an acute dose of 5R produces mutations at a rate linearly proportional to the effects at 1000R and 2000R in the mature gametes of both sexes of *Drosophila*. This formed the basis of linear, non-threshold (LNT) model for radiation risk estimation purposes. The history of the LNT model explains how the idea of a tolerance dose was changed to the linearity model to accommodate possible concerns from the view point of geneticists (Calabrese, 2009).

The LNT model implies the following:

- All doses of radiation, however small, including the environmentally realistic background radiation doses are carcinogenic and mutagenic.

- Radiation doses are cumulative in producing biological effects; that is to say the magnitude of the biological effects (cancer, mutations) will be the same whether a dose is given acutely (say 1 Gy given in one minute) or chronically (say 1 Gy given over a period of 100 days at 0.01Gy per day).

- The effects other than those observed at high doses will not occur at low doses.

During the 1950s and 1960s, the basic mechanisms of the molecular signaling processes and ensuing cellular events following the absorption of radiation energy in the cells and tissues, was not fully understood. The predominant notion was purely physical in the sense that passage of a single photon or charged particle causes damage to vital target molecules in the cell and results in cancer, mutations and cell death.

With the elucidation of the basic mechanisms in radiobiology, and discovery of phenomena like radiation hormesis, radioadaptive response, DNA repair, elimination of damaged cells by one of several pathways of cell death (e.g., necrosis, mitotic catastrophe, apoptosis etc.,) and particularly induction of "differential gene expressions" by low and high doses of radiation, the LNT model has come under severe scrutiny (Jayashree *et al.*, 2001). The 'radiation hormesis' (i.e. occurrence of beneficial effects at low doses) at low doses, in particular, rejects the LNT model (Luckey, 1992).

In the following sections, the growing evidence to question the LNT model not only from radiobiological experiments, but also from life span study (LSS) of the atomic bomb survivors and their children for over 60 years, and of Chernobyl nuclear accident victims for over 20 years, is briefly presented. Today, there is polarisation among the radiation health experts on the adoption of LNT model, not only for setting radiation dose limits for protection of civilians and radiation workers, but also for making predictions of the numbers of cancer deaths among the populations exposed to atomic bombs over 67 years ago, to Chernobyl nuclear accident over 25 years ago, and Fukushima nuclear disaster just one year ago. It may be considered that erring on a safer side is a good idea, but in the process, the possible benefits to health (i.e., reduction in the cancer incidence), and environment (i.e., nuclear power reactors do not emit greenhouse gases) should not be denied to humankind and planet Earth, both of which are at cross-roads today. Moreover, the annual exposure limits prescribed for general public by International Council for Radiological Protection (ICRP) is unscientific and unacceptable based on a cost-benefit analysis. For instance, the annual public dose limit for exposure to artificial sources recommended by ICRP[1] is 1 mSv and for workers the limit is 20 mSv averaged over 5 years with no single year exceeding 50 mSv. The annual average exposure to inhabitants in the high level natural radiation in Kerala is 10 mSv and even goes upto three times this value annually. Does this mean that these inhabitants should be evacuated from these regions? Available records show that people have been living here for over 1000 years! Studies have shown that

---

1. ICRP Publication 103: Recommendations of the ICRP (Elsevier, USA, 2007).

the inhabitants in the high level natural radiation areas in Kerala and China are in good health, and there is no increased incidence of cancer.

## Epidemiological Studies and LNT Model

Although epidemiological studies are purely observational, and the observed effects (mostly cancer) could be due to several environmental, dietary and other confounding factors as well as interactions among them (say radon and smoking) (Cohen, 1990), these are of indicative of trends. Kondo (1993), cites several studies in the UK, Japan, US and China showing lack of a positive relationship and indoor radon concentrations. In fact, lung cancer incidence is reported to be significantly reduced with increase in radon concentrations $\geq 8.0$ pCi/L air. In radiation epidemiology, the cohort members are drawn from those with medical, nuclear accident (Chernobyl) and atomic bomb radiation exposures. Late effects (cancer) are followed in the exposed groups in Life Span Study (LSS). The children of the exposed survivors are subject to genetic investigations for heritable changes induced in their exposed parents. These are briefly as follows.

### Health Effects Attributable to Atomic Bomb Radiations

The Radiation Effects Research Foundation (RERF) is a US-Japanese bi-national body set up in April 1975. This is a successor to the Atomic Bomb Causality Commission (ABCC) set up in 1946 by the then President of the United States. For over 65 years now, the RERF has developed an extremely valuable and scientifically sound database on the 'late' effects (mainly cancer) of radiation on humans. Assessment of the radiation effects directly on humans has revealed that extrapolations of the genetic effects of radiation on the basis of earlier experimental studies with *Drosophila* and mice are inappropriate. Not only are these organisms ephemeral, but they also differ in physiology, gestation, numbers of offspring conceived and delivered etc. (Neel, 1998). Under these circumstances, it is largely because of the age and gender characteristics of the survivor population and the wide range of generally uniform whole body doses that they received, including large numbers with low and moderate doses that the Japanese atomic bomb survivor LSS cohort is regarded as the epidemiological 'gold standard' for assessing radiation health effects in the humans. The LSS is one of the largest and most comprehensive prospective cohort studies ever conducted, with information available on other risk factors such as smoking (Pierce *et al.*, 2003) and diet (Sauvaget *et al.*, 2004).

The LSS cohort includes 120,321 people including 93741 survivors, who were within 10 km of the bomb hypocenter, and also 26,580 who were not in Hiroshima or Nagasaki at the time of bombings, the so-called 'not-in city (NIC)' group. It has now been ascertained (Kodama *et al.*, 2012) that the doses received by individual cohort members range from 'no exposure' to more than 4 Gy, as determined by the latest Dosimetry System 2002 (DS02).

The RERF study also included a population of about 3,000 persons exposed to the bombings, while they were in their mother's wombs (*in utero*). Although by early 1950s it had been known that there was no evidence suggesting genetic changes in the children born later to survivors (i.e., those not exposed in the womb), researchers recognised the need for continued follow-up on children of survivors and established a study population known as the F1 sample. The F1 study comprised 77,000 children, of which about 30,000 have atleast one parent who received a dose greater than 0.005 Sv.

## (i) Heritable Genetic Effects in the Children of Exposed Parents

The RERF studies have led to the conclusion that atomic bomb radiation has not resulted in genetic changes in the children of the parents exposed. Data presented in Table 5.1 are from Kondo (1993).

The absence of the expected and greatly feared genetic and teratogenic effects in the children of the highly exposed atomic bomb survivors is described in a simple and sophisticated statement of Neel *et al.* (1990) as follows:

The children of the most highly irradiated population in the world's history provide no statistically significant evidence that mutations were produced in their parents. In particular, the studies should prove reassurance to that considerable group of exposed Japanese and their children, without whose significant cooperation these studies would have been impossible and who have over the years been subjected to a barrage of exaggerations concerning the genetic risks involved.

These conclusions were subsequently confirmed by elegant molecular biology techniques. No genetic instability of tandem repetitive elements in the germ cells of the atomic bomb survivors was observed (Kodaria *et al.*, 1995).

**Table 5.1**

*Genetic Effects of Radiation in Children of Atomic Bomb Survivors in Hiroshima and Nagasaki*

| Indicator | Frequency (No. Abnormal/ No. Studied) | | Parental Dose[a] | References |
| | Control | Exposed | (rem) | |
| --- | --- | --- | --- | --- |
| Untoward pregnancy outcome[b] | 4.99% (2,257/45,234) | 5.00% (503/10,069) | 36 | (12) |
| Deaths of liveborn children | 7.35% (2,451/33,361) | 7.08% (989/13,969) | 40 | (13) |
| Stable chromosomal aberrations | 0.31% (25/7,976) | 0.22% (18/8,322) | 60 | (14) |
| Aneuploidy | 0.30% (24/7,976) | 0.23% (19/8,322) | 60 | (14) |
| Mutations in blood proteins | $6.4 \times 10^{-6}$ ($3/4.7 \times 10^5$) | $4.5 \times 10^{-6}$ ($3/6.7 \times 10^5$) | 41 | (15) |
| Leukemia | 0.05% (21/41,069) | 0.05% (16/31,159) | 43 | (13) |

*Note*:  a  Sum of average doses to mothers and to fathers.

   b  Congenital malformations, still birthsand deaths in the first 14 days of life.

Data on genetic changes induced in germ-line cells of atomic bomb survivors suggest a mutation-rate doubling dose of 340-450 Rem (i.e., 3.4 to 4.5 Gy) in the humans and this is three to four times higher than doubling dose calculated on the basis of experimental data of spermatogonial mutations in mice (Russell and Kelly, 1982). As already mentioned in section 2-A, there are wide differences between humans and mice in their reproductive structure and physiology. The fact, however, from the present context is that ionising radiation is not mutagenic to humans. This means that if radiation has induced genetic changes in the germ cells, these are not transmitted to zygotes. It should be noted that mutations in germ-line cells are more readily eliminated (i.e., apoptosis, and sieving of defective sperms during fertilisation, zygotic death etc.), whereas the somatic cell mutations are eliminated by a fewer pathways. For these reasons, for over five to six decades, the stochastic effects in all situations of human exposures to ionising radiation mainly refer to cancer morbidity and mortality. Cancer results from mutations in somatic cells.

## (ii) Cancers in the Exposed Survivors

As of 1990, there were 176 leukaemia deaths among 51,114 LSS survivors with a bone marrow dose in excess of 0.005 Sv. This group had 87 deaths beyond expected deaths from leukaemia, which means that 49 per cent of the cases were attributable to radiation.[2] More recently, Kodama *et al.* (2012) have presented an up-to-date report made in 2009 on leukaemia mortality risk. This report states that the estimated attributable fraction of leukaemia deaths among those survivors exposed to >0.005 Gy was 34 per cent. During the period of roughly twenty years (1990-2009), the radiogenic leukaemia among the A-bomb survivors has decreased by about 15 per cent, notwithstanding the fact that the survivors have also grown older by about 20 years.

Radiation link to solid cancers has also been investigated in the LSS survivors. For the average radiation dose of survivors within 2,500 meters (~200 mSv), the increase was about 10 per cent above age-specific rates.[3] More recent analyses (Kodama *et al.*, 2012) suggest that as of 2007, 17,448 solid cancers were identified among the LSS cohort members, of which—for those members with colon doses in excess of 0.005 Gy—about 850 (~11%) of the cases were estimated to be excess cases associated with atomic bomb radiation exposure. The increase in solid cancers is most notably concentrated in breast, thyroid and lung (Davis, 2012). The effect of smoking interacting with radiation in the LSS lung cancer incidence was found quite significant (Furukawa *et al.*, 2010).

---

2. National Research Council Committee to Assess Health Risks from Exposure to Low Levels of Ionising Radiation 2006 Health Risks from Exposure to Low Levels of Ionising Radiation: BEIR VII: Phase 2, Washington, DC: National Academic Press.

3. Ibid.

Another analysis of cancer mortality among Japanese nuclear workers for the period from 1991 to 2002 estimates the excess relative risk (ERR) per radiation dose (Akila and Mizuno, 2012). The estimate is 1.26 ERR/Sv with average individual cumulative dose of 12.2 mSv (<10 mSv,75.4%; 100+ mSv,2.6%). Confounding by alcohol consumption was important since the ERR/Sv of alcohol-related cancer was 4.64 and the ERR/Sv estimate of all cancers excluding leukaemia and alcohol related cancer was 0.20.

The LSS cohort survivors study clearly establishes a radiation link to enhancing the incidence of both leukaemia and solid cancers over the spontaneous rate of increase with ageing. Smoking particularly enhances the occurrence of lung cancers. The question, however, is about the incidence of neoplasmic diseases among humans exposed to low to very low doses. There are two different estimations of cancer incidence based on the LSS cohort survivors data—one by Kondo (1993) and another by Henriksen and Maillie (2003). Kondo (1993) arrives at an apparent threshold of 36 rad (0.36 Gy) for leukaemia, 54 rad (0.54 Gy) for stomach and 50 rad (0.50 Gy) for breast cancers. Henriksen and Maillie (2003) have presented data (Table 5.2), which provide a threshold dose of 100 mSv (10 rem) for radiation-induced increase in solid cancers, and a threshold dose of 200 mSv (20 rem) for leukaemia for the LSS cohort survivors (1950-1990).

**Table 5.2**

*Health Status of the Survivors of A-Bomb (Hiroshima and Nagasaki)*

| Dose/MSv | Number | Leukaemia Deaths | Risk per 10,000 | Number | Solid Cancer Deaths | Risk per 10,000 |
|---|---|---|---|---|---|---|
| 0-5 | 35458 | 73 (64) | 3±3 | 39507 | 4270 (4268) | 0±20 |
| 5-100 | 32915 | 59 (62) | -1±3 | 29960 | 3387 (3343) | 15±20 |
| 100-200 | 5613 | 11 (11) | 0±10 | 5949 | 732 (691) | 70±45* (100m Sv) |
| 200-500 | 6342 | 27 (12) | 24±10* (200 m Sv) | 6380 | 815 (716) | 155±45 |
| 500-1000 | 3425 | 23 (7) | 46±16 | 3426 | 378 (262) | 340±60 |
| 1000-2000 | 1914 | 26 (4) | 120±30 | 1764 | 326 (213) | 640±100 |
| >2000 | 905 | 20 (2) | 310±60 | 625 | 114 (58) | 900±170 |

*Note:* Numbers given within the bracket are the numbers of cancer deaths expected from similar populations not exposed to radiation.

*Source:* Henriksen and Maillie (2003). Data from 1950 to 1990.

Tanooka (2011) recognises the existence of non-tumour doses (i.e., $D_n t$, defined as the highest dose of radiation at which no statistically significant tumour increase is observed over the control level) and attributes it to a function of dose-rate. The dose-rate of ionising radiation that humans have been exposed to, varies over a wide range from 10-9 Gy/min in natural radiation to 107 Gy/min in nuclear bombs explosion and nuclear accidents. The atomic bomb explosions involved very high dose rate exposures for shorter periods. Unlike in the case of Chernobyl nuclear accident, the radioactivity in Hiroshima and Nagasaki was over 90 per cent depleted by one week after the bombings and was even less than background within one year.[4]

Having dealt at length with the LSS cohort survivors of atomic bombs, the major question that needs to be unequivocally answered is, "what is the level of cancer risk following low dose (< 100 mSv) and/or low dose-rate exposures?" The statistical power in epidemiological studies to answer this question at low doses and low dose-rate exposures is limited; however, experimental radio-biological studies can provide authentic answers. Experiments under controlled conditions preclude confounding factors and also help in unraveling the mechanisms. As of now, derivation of low-dose extrapolation factors from analysis of curvature in the cancer incidence dose response in Japanese atomic bomb survivors (LSS) cohort persons, suggests that a threshold at 60 mSv cannot be ruled out for radiation carcinogenesis in the humans (Little and Muirhead, 2000; Preston *et al.*, 2007). Earlier, Kellerer (2000) had noted that epidemiological studies on the Japanese atomic bomb survivors provide the best estimate of cancer risk over the dose range of 20-250 cGy. The uncertainty is about the linearity from the natural background radiation levels up to 20cGy. Yet, Mullenders *et al.* (2009) interpret that significantly increased cancer risk is observed in atomic bomb survivors exposed to lower doses (5 to 150 mSv) of radiation. The epidemiological data seem amenable to drawing slightly variable interpretations.

---

4. See footnote 2.

*Health Effects Attributable Radiation from*
*Chernobyl Nuclear Accident*

The Chernobyl nuclear accident occurred on the 26th April 1986 during a low-power engineering test of the Unit 4 reactor. Improper, unstable operation of the reactor, which had design flaws, allowed an uncontrollable power surge to occur, resulting in successive steam explosions, which severely damaged the reactor building and completely destroyed the reactor.[5]

The atomic bomb explosions in 1945 in Hiroshima and Nagasaki and the Chernobyl nuclear accident in 1986 as well as Bhopal gas tragedy in 1984, resulted in substantially different magnitudes of causalities within day 1 and during the next four months. These data are presented in table 5.3. The data on the deaths and injuries caused by atomic bombs (1945) and the Chernobyl nuclear accident (1986) are based on those published by Kondo (1993).

**Table 5.3**

*A Comparison of the Numbers of Deaths Caused by Atomic Bombs,*
*Nuclear Accident and MIC Gas Leakage*

| | Deaths within Days | | |
| --- | --- | --- | --- |
| *Site* | *Day 1* | *2-120 days* | *Injures* |
| Hiroshima | 45,000 | 19,000 | 119,000 |
| Nagasaki | 22,000 | 17,000 | 110,000 |
| Chernobyl Nuclear accident | 2 | 29 | Not known |
| Bhopal MIC gas leakage | 3000 (official) 6000-7000 (unofficial) | Many more thousands | Several thousands |

Data presented in Table 5.3 must reassure to people who are protesting against the Kudankulam nuclear reactors, that not only there are no significant late effects (cancer) from radiation, but also there have been far fewer deaths due to Chernobyl nuclear reactor accident, as compared to those of Bhopal gas leakage. A-bombs killed thousands of people not just because of radiation, but also

---

5. International Atomic Energy Agency (IAEA) Summary Report on the Post-accident review meeting of the Chernobyl accident. Safety Series No 75 – INSAG – 1, (IAEA, Vienna, 1986).

because of blast (killing 50 per cent), intense heat (killing 35 per cent), with initial radiation killing 5 per cent of the people in the vicinity. Of course, a bomb is designed to cause maximum number of deaths and most severe damage. People in multi-storeyed building would die in large numbers due to collapse caused by an explosive, rather than by the explosion itself. The point to be emphasised is that a nuclear reactor is not a bomb. Even in the remote possibility of an accident, as severe as that of Chernobyl, it is most certainly not an explosion of the sort of a nuclear weapon. This statement is further strengthened by the data (Table 5.4) that relates exposure dose with number of deaths within the first four months (August 1986) after the Chernobyl accident.

**Table 5.4**

*Outcomes of Radiation Exposure among 237 Emergency Workers in Chernobyl Nuclear Site Hospitalised for Acute Radiation Syndrome (ARS)*

| Number of Patients | Estimated Dose (Gy) | Deaths within Four Months |
|---|---|---|
| 21 | 6-16 (Very Very High) | 20 |
| 21 | 4-6 (High) | 7 |
| 55 | 2-4 (Moderate to high) | 1 |
| 140 | Less than 2 | 0 |
| Total 237 | | Total 28 |

*Note*: Data taken from Agence Pour leenergie nucleaire (*http://www.nea.fr/html/rp/chernobyl/c05.html*)

During 1987-2006, 19 Acute Radiation Syndrome (ARS) survivors died for various reasons (7 deaths due to non-cancer diseases like pulmonary tuberculosis, liver cirrhosis etc.), 1 death due to trauma, 6 deaths due to sudden cardiac arrest and 5 from cancers. The major health consequences from acute exposure to the ARS survivors are skin burns and cataract. A few more cases of solid cancers and leukaemia have been reported. The Chernobyl nuclear accident released a mixture of radionuclides into the air over a period of 10 days. Most of them released in large amounts (in terms of activity) were of short half-life; radionuclides of long half-life were generally released in small amounts. The three major radionuclides released were iodine-131 (half-life 8.04 days), cesium-134 (half-life 2.06 years) and cesium -137 (half-life 30.0 years).[6]

In view of the high degree of prevalence of iodine deficiency in the Belarus and Ukraine regions, it was expected that radioactive iodine would be greatly absorbed by thyroid, and consequently thyroid cancers would significantly increase over the normal incidence levels. This is what exactly has happened. If only the authorities had administered potassium iodide tablets to the children and adolescents immediately after the nuclear accident, the uptake of radioactive iodine by thyroid would have been substantially reduced. The United Nations Scientific Committee on the Effects of Atomic Radiation (UNSCEAR) Report[7] brings out that the thyroid cancer, which is generally in the range of 2 to 4 cases per million per year, shot up to 20 to 35 per million during 1991-95. As of 2005, there were approximately 6,000 cases of thyroid cancers and 15 thyroid cancer deaths. A most significant finding is that there was no increased incidence of thyroid cancers among children born after 1986, the reason being that the iodine-131 with half-life of just 8 days had substantially decayed.

As of 2005, among those exposed *in utero* and as children, no evidence has been found of a measurable increase in the increase of leukaemia attributable to radiation exposure. The reason is that the exposure doses from the Chernobyl fallout to the general public were generally small.[8]

## Theoretical Projections of Chernobyl Cancer Morbidity and Mortality

Theoretical projections are meant to guide in decision-making on public health resource management. Since, there is a latent period between exposure and appearance of any increased incidence in stochastic effects, various groups have attempted to predict the health impact on populations exposed to very low (almost

---


6. United Nations Scientific Committee on Atomic Radiation 2011 Sources and Effects of Ionising Radiations, UNSCEAR 2008 Report (Scientific Annex D, Health Effects Due to Radiation from the Chernobyl Accident) Vol II (New York: United Nations)

7. Ibid.

8. Ibid.

equivalent to normal background radiation levels) to low doses. It is emphasised that for more than six million residents of the contaminated areas of the former Soviet Union (i.e., those with Cs-137 levels greater than 37 kBq/m2), who were not evacuated, the average dose to thyroid was about 100 mGy, while for about 0.7 per cent of then, the thyroid doses were more than 1000 mGy. The average thyroid dose to pre-school children was some 2-4 times greater than the population average. For the 98 million residents of the whole of Belarusand, Ukarine and 19 Oblasts of the Russian Federation, including the contaminated areas, the average thyroid doses were much lower, about 20 mGy; most about (93%) received thyroid doses of less than 50 mGy. The average thyroid dose to the residents of the other European countries was about 1.3 mGy [30]. As far as whole body doses are concerned, six million residents of the former Soviet Union deemed contaminated received average effective doses of about 9 mSv for the period 1986-2005, whereas for the 98 million people considered in the three republics, the average dose was 1.3 mSv, a third of which was received in 1986. This represents an insignificant increase over the dose due to background radiation over the same period (~50 mSv). About three-quarters of the dose was due to external exposure, the rest being due to internal exposure. However, for about 150,000 people living in the contaminated areas, the effective dose was more than 50 mSv over the 20-year period. For the population of about 500 million in other countries in Europe, the average dose is estimated to have been about 0.3 mSv, over this period.

Epidemiological studies to assess and project the stochastic effect (cancer) require sufficient statistical power to attest to the occurrence of such stochastic effects and hence their attributability to radiation exposure; the level below, which it is intrinsically impossible to detect such effects, depends on the size of the population that is being studied. Boyce (2012) has discussed at length, the uncertainties in studies of low statistical power.

Cancer is a common disease that has not only genetic pre-disposition but also several aetiological factors—both environmental and dietary as well as smoking, drinking etc. Of late, cancer of one or another type, is estimated to affect about 42 out of 100 people in their lifetime (Boice, 2012). Yet, if 100 people received 100 mSv briefly and were followed for life, only one excess cancer would be predicted.[9] So, the epidemiology cannot detect such small increases, had they occurred. However, epidemiological methods may be able to detect relative risks (RRs) perhaps as low as 1.2 to 1.3 i.e. 20-30 per cent relative excesses. But, the RRs of interest following low doses of radiation (1 to 100 mSv) are of the order of 1.001 to 1.04, thus, not much should be anticipated from direct observations at 10 mSv and below. This again necessitates the interpolation of the effects observed at high and moderate doses to those at low doses. What this means is an inevitable reliance on the LNT model, which from the point of view of the new knowledge of radiobiological mechanisms, is highly questionable.

Furthermore, theoretical projections of stochastic (cancer) effects widely vary depending upon several factors not excluding political and/or ideological considerations. For instance, the Greenpeace has estimated that 270,000 cases of cancer are attributable to the Chernobyl nuclear radiation exposure and out of these 93,000 cases would be fatal.[10] The International Physicians for the Prevention of Nuclear War, a group which won the Nobel Peace Prize in 1985, estimates a death toll between 50,000 and 100,000 among emergency workers and those who helped in the clean-up (*http://www/spiegel.de/international/0,1518,411864,00html*). The Union of Concerned Scientists (*http://www/ucsusa.org/news/ press_release/ Chernobyl_ cancer_ death_toll_0536.html*) also estimates the number of cancer deaths attributable to Chernobyl radiation to about 25,000. Such extremely conflicting reports confuse the general public and enhance the trauma of the exposed survivors. Further, it blocks the avenues for harnessing atoms for peace in nuclear power generation, medicine and agriculture. So, in 2003, eight bodies of the United Nations family (FAO, IAEA, UNSCEAR, UNEP, WHO, World Bank, UNDP and Office for the Coordination of Humanitarian Affairs) and the three republics affected by Chernobyl nuclear accident, launched the 'Chernobyl Forum' to generate "authorative consensual statements" on the environmental

---

9. See footnote 2.

10. See footnote 6.

and health consequences attributable to radiation exposure and to provide advice on issues such as environmental remediation, special health-care programmes and research activities.

Contrary to the very high numbers of cancers (leukaemia + solid cancers) predicted by Greenpeace and a few other above said groups, the cancer effects of radiation exposure from the Chernobyl accident as of 2005, have been fewer. The major cancer incidence was the thyroid cancer. This report estimates about 700 extra leukaemia deaths over life among 5.6 million people, including about 300 among the 600,000 most-exposed liquidators. For the solid cancers, the report estimates about 5,250 extra cancer deaths over life among 5.6 million people, including about 3,650 among the 600,000 most exposed. The report also admits that the statistical power of these studies is low and also there are methodological limitations.[11]

In view of there not being an alarming rise in the stochastic effects attributable to radiation after two decades of the Chernobyl accident, Dr Burton Bennett, Chairman of the Chernobyl Forum (IAEA Press Release on September 05 2005) made the following statement,

> This was a very serious accident, with major health consequences, especially for thousands of workers exposed in the early days who received very high radiation doses, and for the thousands more stricken with thyroid cancer. By and large, however, we have not found profound negative health impacts to the rest of the population in the surrounding areas, nor have we found widespread contamination that would continue to pose a substantial threat to human health, within a few exceptional, restricted areas.

The question, therefore is, how to get more thorough understanding of health effects attributable to low doses of radiation. This comes from experimental low dose radiobiology.

---

11. Ibid.

## Radiobiological Effects Induced by Low and High Doses

As of today, it may not be prudent to conclude that radiation-induced molecular and cell biological events unequivocally establish that low doses are absolutely beneficial. However, what can be most certainly concluded is that radiation-induced molecular signaling at low and high doses are quite different. This in itself constitutes a very serious challenge to the LNT model. This, in fact, should not come as a surprise to those who understand evolution and adaptation of life to much higher doses of ultraviolet and cosmic radiations in the early history of the planet Earth and its living organisms. Our planet is about 4.8 billion years old, and it is believed that the ancestral cell with all the attributes of life was formed about 3.5 billion years ago. Further, there is geological evidence to suggest that there was more than a billion year delay between the rise of photosynthetic cyanobacteria (the first organisms to release oxygen by splitting water molecules in the photosynthetic process) and the time that high oxygen levels began to accumulate in the atmosphere (Alberts *et al.*, 1989). Hence, our ancestral organisms that evolved during the first 1.5 billion years in the earth's primordial atmosphere, deficient in oxygen and ozone, should have been heavily bombarded with ultraviolet as well as other high energetic particulate and electromagnetic radiations of substantially higher doses than these are at present. Life was not extinguished then; in fact, the evolution from prokaryotes to uni- and multicellular eukaryotes continued. Initially, when oxygen concentration in the atmosphere started increasing, it was 'toxic' to several forms of life! Yet, a few organisms developed excellent cellular metabolic and biochemical mechanisms to make effective use of the oxygen through oxidative phosphorylation. However, it should be kept in view that no benefit comes without a cost and in this case, it came in the form of reactive oxygen species (ROS) in the aerobic organisms. The ROS are essential, but could be damaging too as oxygen! Initially, when organisms are young, the ROS-mediated damage is effectively repaired/eliminated, but with ageing, the repair processes become progressively slower and inadequate. Several diseases such as cancer, arthritis, myocardial infarction (Bell *et al.*, 1990) are some of the free radicals-mediated diseases. The law of life is a trinity (birth→growth and reproduction→death). The free radicals reacting with molecular oxygen and forming the ROS, play a significant role in each phase of the trinity. Furthermore, oxygen enhances the radiobiological damage caused by low LET radiations such as gamma-rays and X-rays. This evolutionary history of life on planet also provides a clue as to why bacteria, butterflies and cockroaches, which appeared on the earth hundreds of millions of years prior to the humans, are far more radio-resistant. It is possibly because these organisms developed adaptation to much higher levels of natural background radiation that existed several million years ago, than about 2 million years ago when the closest ancestors of Homo sapiens emerged. It is possible that differences that once existed in the levels of natural background radiation at the time of origin of the different species, are manifested today as variations in their radiosensitivities. It is suggested that these could represent adaptations to different natural levels of radiation. Unfortunately, there have been no systematic investigations to verify these. The ancient wisdom in the statement of Paracelsus (1493-1541), a Renaissance Physician, naturalist and philosopher is evident from the question he posed, "what is it that is not poison? All things are poison and none without poison. Only the dose determines that a thing is not poison."

A number of studies in the literature suggest that low doses are not harmful, but could be even beneficial. Some of these are considered below:

### Effects of Daily Small Dose Irradiation on Life-Span and Cancer Incidence in Mice

After the first experiments of Lorenz *et al.* (1955) showing a significant increase in the life-span of the male mice, daily irradiated with 0.11cGy gamma-rays, several other animal studies have also reported either an absence of adverse effect or even a beneficial (also called hormetic) effect using single or protracted very low doses of radiation. Caratero *et al.* (1998) showed that the female mice continuously irradiated for its whole life with dose-rates as low as 7 or 14 cGy year$^{-1}$ gamma rays had significantly higher mean life-span when compared with controls living in the same room (549±9 days for the control and 673±13 days for both the irradiated groups). A recent paper (Courtade *et al.*, 2002), reports that female mice exposed to 10 cGy year$^{-1}$, using thorium nitrate as the source, had no adverse effects. There was no increased incidence of cancer. Thorium–232 emits gamma rays of 60 Kev and alpha rays of 4 Mev.

### Studies on the Inhabitants in High-level Natural Radiation Areas

With regard to human health effects of high level, natural radiations at low dose-rates, studies of the incidence of cancer and cytogenetic (i.e., clastogenic) effects in the populations inhabiting in the world's high-level natural background areas such as Tongyou of Dong-anling regions in Yangjiang country of China and the Chavara region of Kerala in the Southwest coast of India, are of interest.

In the Chinese studies, the annual effective external dose equivalents (rem) in the High-level Natural Radiation Areas (HNRA) and Control Areas (CA) are 0.210 and 0.077, respectively. Wei (1996) conducted extensive cancer epidemiological studies involving 74,000 people in the HNRA and 77,000 people in the CA during 1970-1986. Their data showed that there were 467 cancer deaths in the HNRA for about 1 million person-years, with an adjusted mortality rate of 48.8/10$^5$ and 533 in the CA for 995,000 person-years, with an adjusted mortality rate of 51.1/10$^5$. The difference was not statistically significant. If, however, the comparison is made for cancers other than leukaemia among the inhabitants in the age group of 40 to 70 years, the cancer mortality rate is significantly lower in the HNRA. There was an age-dependent increase in unstable chromosomal aberrations (i.e., chromosomal translocations and inversions) in the inhabitants of both HNRA and CA of China (Jiang *et al.*, 1996). That is as to be expected.

The high-level natural radiation areas (HNRA) in Kerala have deposits of radioactive monazite-bearing sand. This area is densely populated and has been inhabited for over 1000 years. The radiation levels are high, varying from 1.0 to over 35.0 mGy/year in different areas. Radiation exposure is due to thorium and its decay products emitting alpha-and beta-particles as well as gamma radiation; the radiation exposure to the inhabitants occurs both externally and internally by inhalation and ingestion. Since 1986, the Bio-medical group of the Bhabha Atomic Research Centre, Mumbai, has been assessing the cytogenetic effects in the new born population from high-level natural radiation areas. The results are published (Cherian *et al.*, 1999; Jaikrishan *et al.*, 1999). The areas with a dose rate of 1.0 to 1.50 mGy/year (average 1.15 mGy/year) was taken as the control areas (CA) and those above 1.50 mGy/year (going upto 35 mGy/year) as high-level natural radiation areas (HNRA). A total of 963,940 metaphases from 10,230 newborns have been screened for various types of chromosomal aberrations. Comparison of 8,493 newborns (804,212 cells) from HNRA and 1,737 newborns (159,728 cells) from CA did not show any significant difference in the frequency of dicentrics, translocations, inversions or other types of aberrations known to be associated with radiation exposure. Their studies did not show any association between the incidence of chromosomal aberrations and the radiation levels, which has a wide, range from 1.0 to over 35 mGy per year. The authors (Cherian *et al.*, 1999) conclude, "Thus, although the average dose-rate increased over 25-fold in the highest exposure group (average 31.41 mGy/year) compared to normal-level areas (average 1.15 mGy/year), there was no significant increase in chromosomal aberrations between the two groups of newborns." They also did not find any significant differences among newborns from HNRA and CA with respect to Down syndrome and other autosomal or sex aneuploids. More recently, it has been reported (Das and Karuppasamy, 2009) that the baseline frequency of micronuclei in HNRA of Kerala newborns is not significantly different from the CA newborns. Both the Chinese and Indian studies have also confirmed that there is no increased incidence of cancers in the HNRA. Extensive studies conducted in Yangjiang of China (Wei, 1996) and the Chavara–Neendakara belt of Kerala (Nair *et al.*, 1999), suggest that there is no difference in the incidence of cancers among the inhabitants of high-level natural background areas. In Kerala, there was, however, a small increase in the incidence of cervical cancer, but the radiogenic breast cancer was slightly reduced in the females. These data seem to suggest confounding factors.

Kesavan (1996) has reviewed the elegant studies conducted since 1965 by Bhabha Atomic Research Centre, Mumbai on the biological effects of high-level natural radiation on plants and rats (*Rattus rattus*) that live in the burrows dug in the radioactive sand (sand containing thorium). These early studies also reported that the mean daily intake of Radon-228 by the population living in the monazite area of Kerala is about 162 pCi, which is about 40 to 50 times the normal daily intake of this radionuclide in the control populations in Kerala (Mistry *et al.*, 1965). It was also known that several plant species, which grow in the monazite sand, uptake considerable quantities of radionuclides. In these plants, chromosomal aberrations were observed. However, these aberrations observed in the root cells are not transmitted via germ cells to the progeny plants. Another study (Nayar *et al.*, 1970) showed that there was no difference in the frequency of somatic mutations in the stamen hairs of *Tradescantia* following exposures at dose-rates of 0.03 mR/hr (control) 0.08 mR/hr, 0.16 mR/hr, 0.32 mR/hr, 0.65 mR/hr) and 1.30 mR/hr for 280 days. The total exposure at the end of 280 days was 0.201 R (0.201 cGy) for the control and 8.736 R (8.736 cGy) for the maximally exposed group. The mutation frequencies in the sensitive stamen hairs system were $2.77 \pm 0.706$ and $3.63 \pm 0.819$ for the control and the irradiated groups, respectively. The observed differences are not statistically significant. Several other papers on radiation dose and health effects are found elsewhere.[12]

Radiation Hormesis

Jayashree *et al.* (2001) have noted in their review that benefits derived from low doses of toxic substances have been reported over many centuries. The term used to describe the beneficial effects at low concentrations/doses of substances/agents, which at high doses are toxic, is 'hormesis' (Greek 'Hormoligosis'). Hormesis means, 'to excite in small amounts'. Hormesis can be defined as any physiological effect that occurs at low doses and which cannot be anticipated by extrapolating from toxic effects noted at high doses.

---

12. *High Levels of Natural Radiation: Radiation Dose and Health Effects* (Editors Luxin W, Sugahara T and Tao Z) Elsevier 1997.

Luckey (Luckey, 1992) in his book, *Radiation Hormesis* presents a comprehensive list of studies by several authors, showing that low doses of ionising radiation produce beneficial effects in organisms and are 'hormetic'. For example, some degree of exposure to sunlight is necessary for the UV to produce vitamin D in the skin, but excessive exposure could cause skin cancer. One cannot lead a healthy life by completely avoiding sunlight, which is composed of electro-magnetic radiation of widely different wavelengths. An experiment conducted by Planel *et al.* (1987) showed that cell proliferation of a protozoan—*Paramecium tetraurelia* was significantly reduced in sub-ambient levels of ionising radiation. The first study of the biological effects of significantly reduced background radiation levels of ionising radiation was performed in a room~2500 m below the ground in the middle of Simplon tunnel (Luckey, 1992). In this experiment, the cosmic radiation was reduced to about 1 per cent with energies less than 5 Mev. A 10 cm thick iron-box within the tunnel further reduced the earth radiation to about one-tenths. The viability of the eggs of *Artemia*, a brine shrimp, in the sub-ambient radiation environment decreased rapidly to about 10 per cent, whereas the controls (i.e., those placed in normal radiation levels reaching the earth) maintained a hatching rate of >60 per cent during the experimental period.

In another study, (Yamashita *et al.*, 1991) it was found that the rate of division of mammalian cells in culture was significantly depressed when the Petri dishes containing the cells were kept inside of a lead-shield. However, when a small amount of cobalt-60 (gamma– rays) was placed inside of the lead chamber, the mitotic cell division rate was significantly restored.

There are several studies, which show hormetic effects of low doses of radiation. This is contrary to the expectation based on the LNT model.

The LNT, threshold and hormetic models are represented as in Figure 5.2.

**Figure 5.2**

*Three Hypothetical Models for Dose Response Curves*

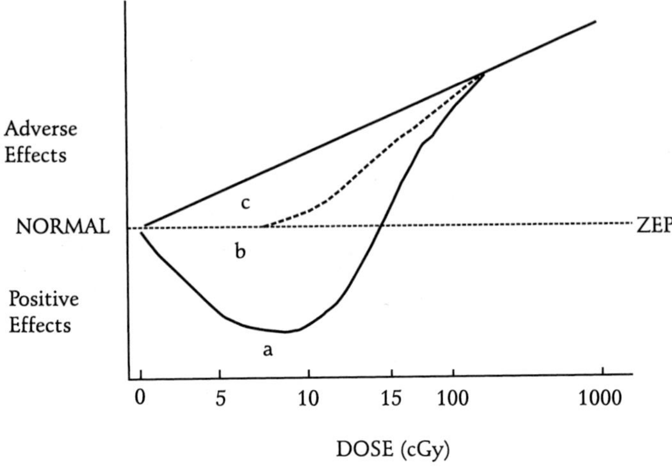


*Note*:  a: hormesis; b: threshold; c: linear, no-threshold (LNT); ZEP: Zero Equivalent Point

Further, in view of the above mentioned observations that sub-ambient/sub-optimal radiation environment causes debilitation and morbidity, a 'complete dose-response curve'—(one that considers the debilitating physiological effects in organisms conditioned under sub-ambient radiation environment) is proposed (Luckey, 1992) as shown in Figure 5.3.

**Figure 5.3**

*Complete Dose Response Curve*

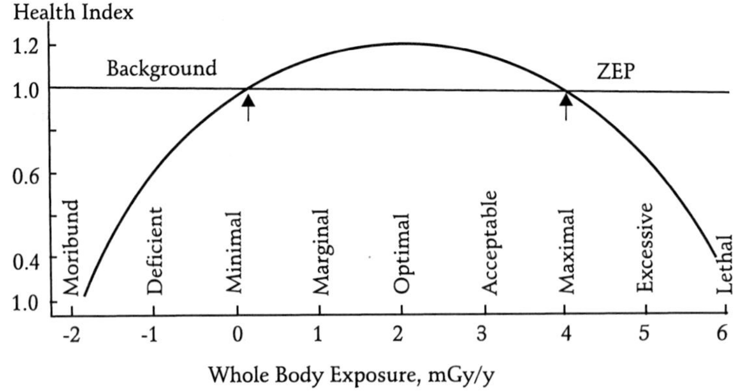


*Note*:  Low background radiation also results in adverse health effects.

Recently, Zhang *et al.* (2010) have shown that bone marrow cells exposed to low doses (6cGy and 8cGy) and infused into heavily irradiated (7.5 Gy) mice, greatly facilitate haematoposetic reconstitution of the heavily-irradiated recipient mice. Nowosielska *et al.* (2011) have reported anti-neoplastic and immunostimulatory effects of low whole-body irradiated (0.01,0.02 or 0.1 Gy of X-rays/day for five days/week for two weeks) mice. The observed suppression of the growth of pulmonary tumour colonies in the irradiated mice is attributed to stimulation of anti-tumour reactions mediated by NK cells and/or cytotoxic macrophages.

## Radioadaptive Response

Contrary to the general expectation that radiation doses act in a cumulative manner (i.e. a dose of 0.1 cGy given first and another dose of 0.9 cGy given later would produce a biological effect equivalent to 0.1 cGy + 0.9 cGy = 1.0 cGy), a lower dose given first often protects the cells and organisms against a subsequent high dose exposure. This is the phenomenon of radioadaptation. The low doses that are effective in inducing radioadaptive responses are known as priming doses. Radioadaptive responses have been observed in multiple biological end-points such as unscheduled DNA synthesis, micronuclei, chromosomal aberrations, gene mutations and cell survival. Jayashree *et al.*, 2001 have compiled a few of the earlier reports of radioadaptation. More recently, Lu *et al.* (2009) have reported that 0.1 Gy ionising radiation—induced adaptive response substantially reduces the mutation frequencies induced by high doses of ionising radiation and UV in mammalian cell lines. The most interesting studies are the chromosomal sensitivity of temporary nuclear workers and adaptive response induced by occupational exposures. The Micronucleus (MN) frequencies following occupational exposures up to 10 mSv are not significantly different from the unirradiated (control) persons. For workers who received the highest doses (up to 10 mSV), a statistically significant (p<0.05) decrease of the *in vitro* induced MN yields and increase of the dose-rate sparing was observed (Thierens *et al.*, 2002). Low doses of diagnostic energy of X-rays, protect against neoplastic transformation in *in vitro* (Redpath *et al.*, 2003). In this experiment, low doses (< 1 cGy) of 60 kVp X-rays protected He La X skin fibroblast human hybrid cells against neoplastic transformation.

## Bystander Effects and Genomic Instability

When cells are irradiated in vitro, it is generally expected that cells in the population that received no direct radiation exposure would not have any genetic and chromosomal damage. In the last ten years, however, evidence has been presented indicating that genetic changes such as increased levels of sister chromatid exchanges (Nagasawa and Little, 1992), mutations (Zhou *et al.*, 2000), micronuclei (Hickman *et al.*, 1994) and DNA damage-inducible proteins (Azzam *et al.*, 1998) occur in a greater-than-expected number of cells in cultures exposed to very low fluences in which only a small fraction of the cells are actually traversed by a particle track and thus, directly exposed to radiation. These observations suggest that cells not themselves directly traversed by ionising radiation behave as if they have been hit. This phenomenon is known as the 'bystander effect'. Mitchell *et al.* (2004) define bystander effect as the observation of a biological response in cells that are not themselves traversed by ionising radiation, but which communicate with cells that are. Several aspects of this phenomenon are found in a recent review (Morgan, 2003: 567, 581).

The first and the foremost question is whether bystander effect (mostly noted when the cells are exposed to high LET alpha-rays and not low LET gamma rays), is in contrast to the adaptive response where a low priming dose of irradiation (<10 cGy) induces a protective adaptive response often against a high challenge dose (Redpath *et al.*, 2001; Broome *et al.*, 2002). Mitchell *et al.* (2004) have shown that an adaptive dose of X-rays cancelled out the majority of the bystander effect produced by alpha-particles. Many radiation researchers studying the bystander effects have addressed the mechanisms by which damage signals may be transmitted from irradiated to non-irradiated bystander cells. There is direct evidence for the involvement of connexin-43 in mediating intercellular communication, in the transmission of damage signals to non-irradiated cells (Azzam *et al.*, 2001). Further, cells that are genetically compromised to perform connexin 43-mediated GJIC do not exhibit bystander response. Although several papers also show that bystander effect is normally observed in cells exposed to alpha-particles (high LET) and not to gamma rays (low LET) (Mitchell *et al.*, 2004; Sowa *et al.*, 2010), and also largely observed in *in vitro* cell cultures and not in *in vivo* systems, the most important question, however, is about its biological significance for radiation protection.

Before discussing the important issue of the biological significance of bystander effects, a related phenomenon, the radiation-induced genomic instability needs to be mentioned. Radiation-induced genomic instability is seen at a high frequency in many cell divisions after the radiation exposure. The instability results in increased frequency of mutations, chromosome aberrations, and cell killing. Further, the heavy metals of relevance to human health also induce genomic instability (Coen *et al.*, 2001).

The magnitude of the response for all of these new phenomena (i.e., bystander effect, genomic instability etc.) has been shown to be dependent on the genetic background of the cells, tissues and organisms in which they are being measured. Research to define the role of inter-individual variability and genetic susceptibility to radiation exposure, is an important focus area for current and future research.

There are two other major view points. One is that occurrences of bystander effect and genomic instability suggest a benign effect that does not require to be eliminated by DNA repair/cell death etc. An established damage to DNA, a manifest gene mutation, and also certain types of chromosomal aberrations may be compatible with the survival of the exposed cell and its progeny. Thus, a mutant cell may survive through several cell divisions if there is no selection against it (Zetterberg *et al.*, 2000). The reference here is to the mutation in the glycophorin—a locus in human erythrocytes, which has no impact on cell survival. On the other hand, when mutations in the HGPRT locus in human lymphocytes were studied in atomic bomb survivors many years after exposure, only weak effects were detected, indicating selection against these mutant cells (Hirai *et al.* (1995). Most reciprocal translocations in lymphocytes persist for several years, indicating limited selection against the affected stem cells. The use of reciprocal translocations as a retrospective biodosimeter for exposure to ionising radiation, is standard practice (Matsumoto *et al.*, 1998). The reciprocal translocations are induced only at moderate to high doses, and not at hormetic (priming) low doses.

The other view now gaining wider consideration is that the radiation-induced bystander effects are actually an adaptive response to low dose exposures (Belyakov *et al.*, 2002; Iyer and Lehnert, 2002; Mothersill and Seymour, 2006). It must also be pointed that there exists a relationship between radiation-induced low-dose hypersensitivity and the bystander effect (Mothersill *et al.*, 2002). The hypersensitivity refers to cell death, that is the elimination of cells, which carry potential genetic changes to result in cancer, from an irradiated population. In nutshell, low doses protect cells and tissues against cancer risk by enhancing one or more of several pathways to their elimination by cell death. While a death response is generally regarded to be an adverse effect, it is indeed an effective protective mechanisms from basic cellular radiobiological point of view. Kondo (1993) in his interesting book, *Health Effects of Low-level Radiation*, presents experimental evidence for 'cell-replacement repair' which is completely error-free than DNA repair, which can also be DNA misrepair. Cell replacement repair involves the elimination by death of the damaged cells as the first step, and then their replacement with normal (undamaged) cells from the stem cell populations.

Cell death occurs in normal situations also in most organisms as a normal event in development and response to cellular damage by environmental and dietary genotoxins. The bystander effects and genomic instability are by no means limited only to low dose ionising radiation; low dose chemical toxicity, (caused by heavy metals etc) also induce genomic instability (Coen *et al.*, 2001; Zetterberg *et al.*, 2000; Hirai *et al.*, 1995); Matsumoto *et al.*, 1998; Belyakov *et al.*, 2002; Iyer and Lehnert, 2002; Mothersill and Seymour, 2006). Furthermore, the observation that repair-deficient cells have larger death-inducing bystander effects than the corresponding repair–proficient parent lines (Mothersill, 2002) suggests that bystander effects and genomic instability are not unique to cells and tissues exposed to ionising radiation, but indeed are universal protection mechanism evolved over millions of years; in this, the damaged cells communicate with the undamaged cells (Bauer, 2007).

It goes beyond the purview of this review to discuss the various types and pathways of death. However, it is briefly stated that in cancer radiotherapy, the goal is killing the cancer cells without, of course, causing death to normal cells as far as possible. The point is that in cancer radiotherapy, cells are killed not by the initial physical and physicochemical events initiated by the ionising radiation, but by the formation of reactive oxygen species (ROS) and these variously signal the expression of several genes leading to one or the other form of cell death. Abend (2003) has discussed several different forms of cell death (terminal differentiation, micronucleation, mitotic catastrophe, necrosis, apoptosis etc). Apoptosis is one of a number of active forms of cell death and is not necessarily p-53 gene-dependent. The last couple of years have witnessed substantially newer understanding such as that high LET heavy ion radiation induces p-53 independent apoptosis (Mori *et al.*, 2009) and that several pro-apoptotic proteins (e.g., BH3: only protein bid) (Maas *et al.*, 2011) also induce apoptosis independent of p-53.

## Differential Gene Expression at Low and High Dose, Immunomodulation, and Induction of Intercellular Apoptosis

It is now becoming increasingly evident that many of the newly recognised effects are similar to systemic stress or innate immune responses, in that there is no simple relationship between exposure and effect and the outcome is not obviously dependent on dose or number of cells hit by radiation (Zhou *et al.*, 2000; Mothersill and Seymour, 2006; Azzam *et al.*, 2003). With an ever-increasing, uncontrolled dumping of electronic waste and with the knowledge that heavy metals also elicit biological response in the form of genomic instability (Coen *et al.*, 2001), it is obvious that it is not specifically induced by ionising radiation. The paper (Mitchell *et al.*, 2004) that an adaptive dose of X-rays (low LET) cancels out the majority of the bystander effects produced by X-rays does not mean that these two phenomena are antagonic to each other. On the other hand, these two represent two different pathways of response towards cellular protection. As of now, it seems reasonable to assume that adaptive response results from both stimulated DNA repair and also from diminished fixation of double-strand breaks (DSBs) in the transcription factories (Radford, 2002). An essential part of the adaptive response is generation or receipt and transmission of a signal that directly induces initiation of a cellular response that dimishes the effects of DNA damage. A hypothesis has been proposed (Szumiel, 2005) that priming stimulus initiates signaling that creates a possibility to switch off transcription of genes damaged as a result of exposure to the challenging dose. That radioadaptation is present in radiosensitive (with DSB repair defects e.g. SCID, Fanconi anaemia etc) cells and the radioresistant cells alike (Sasaki *et al.*, 2002) suggests that priming dose initiates signals to reduce the DSB fixation. These studies, however, need further investigation.

Most interesting observation, however, is the differential influence of low and high doses of ionising radiation on gene expression. Low radiation dose signals are transduced by a protein kinase C (PCK)-p38, mitogen-activated protein kinase (p38 MAPK) —phospholipase C (PLC), signaling pathways. At high doses, ERK (extracellular signal regulated protein kinase), JNK (Jun N–terminal kinase) and Wip 1 (wild-type p-53 induced phosphatase) predominate and negatively regulate the PKC—p38 MAPK-PLC signaling (Szumiel, 2005).

It is becoming increasingly necessary to focus on differential gene expression induced at low and high exposures. In their review, Jayashree *et al.* (2001) have compiled a few reports on this aspect. For instance, Liu *et al.* (1996) have shown that in thymocytes, X-rays induce an increase or decrease in apoptosis depending upon the dose. There is an observation of induction of stress proteins when mammalian cells were exposed to low dose (4 cGy), but not when exposed to 10 cGy of X-rays (Nogami *et al.*, 1993).

Yin *et al.* (2003) have characterised the cellular functions associated with the altered transcript profiles of mouse brain exposed to low-dose (0.1 Gy) and high dose (2 Gy) in vivo gamma-irradiation. Brain irradiation modulated the expression patterns of 1574 genes, of which 855 showed more than 1.5 fold variation. About 30 per cent of genes showed dose-dependent variations, including genes exclusively affected by 0.1 Gy. About 60 per cent of the genes showed time-dependent variation with more genes affected at 30 min than at 4 h. Early changes involved signal transduction, ion regulation and synaptic signaling. Low-dose radiation also modulated the expression of genes involved in stress response, cell-cycle control and DNA synthesis/repair. The authors (Yin *et al.*, 2003) conclude that the doses of 0.1 Gy induced changes in gene expression were qualitatively different from those at 2 Gy. The findings suggest that low dose irradiation of the brain induces the expression of the genes involved in protection, and reparative functions, while down-modulating genes involved in neural signaling activity.

Differential gene expression is not, however, limited only to those involved in cell-division check-points, growth arrest, DNA repair and/or cell death by one of many pathways of which apoptosis is just one. It had already been pointed out that the apoptosis could be either p-53 dependent or independent. The point now being made is that concurrent attention to immuno-modulation at low doses is very essential. Nearly 25 years ago, James and Makinodan (1988) reported on the T cell potentiation in normal and autoimmune-prone mice after extended exposure to low doses of ionising radiation and/or caloric restriction. These observations provide a basis for the reports (Lorenz *et al.*, 1955; Caratero *et al.*, 1998; Courtade *et al.*, 2002) that life-span of mice given daily low dose irradiation is significantly enhanced. There is also a recent report (Ina and Sakai, 2004) that life-span of MRL-lpr/epr mice given chronic low-dose gamma irradiation at 0.35 or 1.2 mGy/h is significantly prolonged. The recent reports on the immuno-modulation by low dose irradiation from the Bio-medical Group of Bhabha Atomic Research Centre, Mumbai, are significant. Dr Sainis and co-workers (Pandey *et al.*, 2005) have demonstrated the low dose stimulation of $CD^+$ T cells. Further, in a subsequent paper, they (Shankar *et al.*, 2006) have studied bystander effects in lymphocytes using filterates from 0.1, 0.5 and 1.0 Gy irradiated lymphocytes. The filterate is referred to as 'irradiated conditioned medium' (ICM). It has been demonstrated that soluble factors released from irradiated lymphocytes initiate a signaling cascade in unirradiated lymphocytes resulting in increased response to mitogen and radio-resistance. Intercellular signaling through reactive oxygen species has been reported as a frequent cause of nontargeted radiation effects (Bauer, 2007; Portess *et al.*, 2007). Portess *et al.* (2007) have shown that low dose irradiation of non-transformed cells stimulates selective removal of precancerous cells via intercellular induction of apoptosis. This process is controlled by TGF-beta and it seems to depend on the induction of peroxidase release. Low dose radiation enhances superoxide anion generation of transformed target cells. Most noteworthy observation, however, is that doses as low as 0.2 mGy alpha particles and 2 mGy of gamma rays of non-transformed cells cause a marked increase in apoptosis induction in transformed cells. The low dose radiation-induced enhancement of intercellular induction of apoptosis primarily involves the HOC/hydroxyl radical and the NO/peroxy nitrite signaling pathway (Bauer, 2007; Portess *et al.*, 2007). They suggest a model in which non-transformed cells hit by low dose radiation respond both by release of effector molecules and TGF-beta. The TGF-beta then triggers neighbouring cells (that had not been hit by radiation) to propagate TGF-beta signaling and effector molecules release. The suggestion is that the removal of transformed cells is enhanced by low dose radiation of non-transformed cells through involvement of TGF-beta and intercellular ROS signaling. Bauer (2007) concludes that low dose radiation-triggered enhancement of apoptosis and autocrine self-destruction might represent a potential control system during carcinogenesis. However, modifications of the complex intercellular ROS-based signaling system may also lead to configurations in which low dose radiation attenuates ROS-mediated apoptosis induction. In such circumstances, it is not known whether another pathway to eliminate the transformed cells would be switched on. While it is known that there are multiple switches for cell death (Abend, 2003), the initiation of specific signals to each of these needs to be elucidated. Therefore, the area for future research is the elucidation of molecular signaling for differential gene expression at low and high doses.

Birth, growth and death are all essentially genes-controlled processes. No dose of a physical or chemical agent can kill a cell or organism without molecular signaling-mediated gene expressions to check cell division for DNA repair, altruistic cell death or carcinogenesis etc. The new knowledge is that low doses induce death more specifically to the DNA-damaged cells so that the potential for carcinogenesis is eliminated. At high doses, this beneficial pathway seems to be somehow obliterated. The question to be answered is whether these radiobiological observations made so far support the LNT model.

## Low Dose Radiation Biology and Cancer Risk Estimation at Very Low Doses

Epidemiological evidence shows convincingly that doses of ionising radiation above about 10-50 mGy enhances the occurrence of cancer (Brenner *et al.*, 2003). However, at doses lower than 10 mGy, even the largest epidemiological studies lack sufficient power to answer the question unequivocally. This situation leads to exactly opposite views on the acceptance of LNT model for risk estimations at low doses. Two expert reports have been published recently, which give diametrically opposite expert opinions. The Biological Effects of Ionising Radiation (BEIR)–VII report, from US National Academy of Sciences,[13] concludes that, at low doses, as the dose is lowered, the cancer risk simply decreases proportionately—a 'linear, no threshold' model down to lowest doses. The French Academy of Sciences (FAS) (Tubiana *et al.*, 2005), does not agree with this. Based on the studies of Rothkamm and Lobrich (Rothkamm and Lobrich, 2003) that the phosphorylated H2AX foci-induced at very low doses in *in vitro* cells are eliminated by apoptosis, the FAS (Tubiana *et al.*, 2005) claims that total elimination of cells carrying DSBs is far better than subjecting them to repair that also has a certain probability of misrepair. Cells with misrepaired DNA can initiate carcinogenesis. Brenner (2007) points that Lobrich *et al.* (2005) in

---

13. See footnote 2.

a subsequent paper have shown that under the in vivo conditions, however, the same standard DSB repair processes (which can also result in misrepair) are operative in lymphocytes receiving doses from 3 mGy to 30 mGy. Hence, the point is that it is hard to assert that cancer risks are really zero at very low doses. It is, of course, true that intercellular signaling could well modify the shape of the dose-risk relation away from linearity at very low doses, but the magnitude or even the direction of such modifications away from linearity are, as yet, unknown. What is, however, now well established is that multiple mechanisms are possibly initiated in a dose-dependent fashion. It is also now clear that reactive oxygen species (hydroxyl radicals, hydroperoxides, superoxide anion etc) do not directly inactivate the cells by abstraction of electrons from the DNA, membranes and other macromolecules but by variously inducing the molecular signaling for gene expressions depending on the initial dose of irradiation. At higher doses, the increased magnitude of damage to a large population of cells probably overwhelm the coping capacity to repair/eliminate the damage through mechanisms evolved over millions of years. This will be an exciting area for high level research in the near future.

## Emerging Evidence of High Level Research

Just after submission of this paper to Indian National Science Academy in March 2012, the author got access to a recent paper (Neumaier *et al.*, 2012) presenting evidence for formation of DNA repair centers and dose-response non-linearity in human cells. Traditionally, it was assumed that the repair proteins (such as p53 binding protein, 53BPI) would be recruited to the location of the DSBs to repair/eliminate the DSBs. Using live imaging and mathematical fitting of Radiation-Inducing Foci (RIF) kinetics, Neumaier *et al.* (2012) show that multiple DSBs about 1 to 2 $\mu$m apart rapidly cluster into repair centers, which also explains the observation of an absolute RIF yield that is surprisingly smaller at higher doses: 15 RIF/ Gy after 2 Gy exposure compared to 64 RIF after 0.1 Gy. This discovery of DSB clustering over such long distances casts considerable doubts on the assumption that, risk to ionising radiation is proportional to dose, and instead provides a mechanism that more accurately addresses risk dose dependency of ionising radiation. The authors are of the view that their findings suggest that extrapolating risk linearly from high dose as done with LNT could lead to overestimation of cancer risk at low doses.

The LNT controversy can be settled mostly with the help of cell and molecular biological studies oriented towards DSBs repair and death of damaged cells at low and high doses. And even more basically, studies on the differential gene expression at low and high doses should be supported.

## Conclusions

Thomas Kuhn in 1962 noted that paradigm shift in science is very difficult. Despite compelling evidence against a prevailing concept, a change for a more appropriate one is often strongly resisted. This indeed is true for the LNT model. Several radiobiological phenomena such as radiation hormesis, radioadaptive response, adaptive bystander effects, low-dose hypersensitivity in respect of cell-killing etc., challenge the LNT model. In particular, the differential gene expression at low and high doses resulting in opposite effects (i.e., beneficial at low and harmful at high doses) rejects the LNT hypothesis. Further, the radiation adaptation induced by low doses (i.e., priming doses) observed in several experimental studies reveals, how, in the course of evolution and natural selection, organisms have become adapted to natural level background radiation, and adaptation to still higher levels of radiation is just natural.

In the above mentioned context, Allison's (2009) book *Radiation and Reason* and Don Luckey's (1992) book *Radiation Hormesis* build a firm foundation to bring about the necessary paradigm shift with regard to making radiation risk estimation at low doses based on the LNT model.

This review paper concludes that there is considerable scientific evidence to argue that low and high doses produce exactly opposite biological effects, and therefore, the LNT hypothesis is incorrect.

# References


Abend, M. (2003). *Int J Radiat Biol* 79: 927.

Akila, S. and S. Mizuno J. Radiol (2012). *Prot* 32: 73.

Alberts, B., D. Bray, J. Lewis, M. Raff, K. Roberts and J.D. Watson (1989). *Molecular Biologyof the Cell*, 2nd (ed.), Garland Publishing, New York

Allison, W. (2009). *Radiation and Reason:* The Impact of Science on a Culture of Fear, York Publishing Services, York UK, p. 216.

Awa, A.A. , T. Honda, S. Neriishi, T. Sofuni, H. Shimba, K. Ohtaki, M. Nakano, Y. Kodama, M. Itoh and H.B. Hamilton (1989). *Cytogenetic Study of the Offspring of Atomic Bomb Survivors, Hiroshima and Nagasaki* (RERF TR 21-88), Radiation Effects Research Foundation, Hiroshima p.1.

Azzam, E.I. , S.M. Toledo and J.B. Little (2001). *Proc Natl Acad Sci* 98: 473. USA.

———. (2003). *Oncogene* 13: 7050.

Azzam, E..I, S.M. de Toledo, T Gooding and J B Little (1998). *Radiat Res* 150: 497.

Bauer, G (2007). *Int J Radiat Biol* 83: 873.

Bell, D., M. Jackson, J.J. Nicoll, A. Millar, J. Dawes and A.L. Muir (1990). *Br Heart J* 63: 82.

Belyakov, O.V., M. Folkard, C. Mothersill, K.M. Prise and B.D. Michael (2002). *Radiat Prot Dosimetry* 99: 249.

Boice, J.D. Jr. (2012). *J Radiol Prot* 32, N33.

Brenner, D.J. (2007). *Int J Radiat Biol* 83: 493 In: Environmental Mutagen Society Symposium on "Risks of Low Dose, Low Dose Rate Exposures of Ionising Radiation to Humans" (Chairs: (Morgan W F and Schwartz J L) 491

Brenner, D.J., R. Doll, D.T. Goodhead, D.J. Hall, C.E. Land, J.B. Little, J.H. Lubin, D.L. Preston, R.L. Preston, J.S. Puskin, E. Ron, R.K. Sachs, J.M. Samet, R.B. Setlow and M. Zaider (2003). *Proc Natl Acad Sci* 100: 13761. USA.

Broome, E.J., D.L. Brown, and R.E. Mitchell (2002). *Radiat Res* 158: 181.

Calabrese, E.K. (2009). *Arch. Toxicol* 83.

Caratero, A., M. Courtade, L. Bonnet, H. Planel and C. Caratero (1998). *Gerontology* 44: 272.

Cardis, E. *et al.* (2005). *J. Natl. Cancer Inst* 97: 724.

Cherian, V.D., C.J. Kurien, B. Das, E.N. Ramachandran, C.V. Karuppasamy, M.V. Thampi, K. P. George, P.C. Kesavan, P.K.M. Koya and P.S. Chauhan (1999). *Radiat Res* 152. S pp 154.

Coen, N., S. Mothersill, M. Kadhim, E.G. Wright (2001). *J. Pathol* 195: 293.

Cohen, B.L. (1990). *Environ. Res* 53: 193

Courtade, M., C. Billotet, G. Gasset, A. Caratero, J.P. Charlet, B. Pipy and C. Caratero (2002). *Int J Radiat Biol* 78: 845.

Das, B. and C.V. Karuppasamy (2009). *Int J Radiat Biol* 85: 272.

Davis S (2012). *J Radiol Prot* 32. N 21.

D.L. Preston, E. Ron, S. Tokuoka, S. Funamoto, M. Nishi, M. Soda, K. Mabuchi and K. Kodama (2007). *Radiat Res* 168: 1.

E. Yin, D.O. Nelson, M.A. Coleman, L.E. Peterson and A.J. Wyrobek (2003). *Int J Radiat Biol* 79: 759.

Furukawa, K, L Preston, S Lonn, S Funamoto, S Yonehara, T Matsuo, H Egawa, S Tokuoka, K Ozasa, F Kasagi, K Kodama and K Mabuchi (2010). *Radiat Res* 174: 72.

Glass, H.B. and R. Ritterhoff (1961). *Science* 133: 1366.

Henriksen. T and H.D. Maillie (2003). *Radiation and Health.* Taylor and Francis, London

Hickman, A.W., R.J. Jaramillo, J.F. Lechner and N.F. Johnson (1994). *Cancer Res* 54: 5797.

Hirai, Y., Y. Kusunoki, S. Kyoizumi, A.A. Awa, D.J. Pawel, N. Nakamura and M. Akiyama (1995). *Mutat Res* 329: 183.

Ina, Y. and K. Sakai (2004). *Radiat Res* 161: 168.

Jaikrishan G., V.J. Andrews, M.V. Thampi, P.K.M. Koya, V.K. Raja and P.S. Chauhan (1999). *Radiat Res* 152 (Suppl) S 149.

James, S.J. and T. Makinodan (1988). *Int J Radiat Biol* 53: 137.

Jayashree, B., J.P.A. Devasagayam and P.C Kesavan(2001). *Curr. Science* 80: 515.

Jiang, T., Y. Wanga, D.Q. Chen, Y.R. Yuan, L.X. Wei, H. Hayata, H. Morishima, S. Nakai and T. Sugahara (1997). *HighLevels of Natural Radiation 1996* (eds.) L. Wei, T. Sugahara and Z. Tao) Elsevier Science BV (1997) Amsterdam

Iyer, R. and B.E. Lehnert (2002). *Mutat Res* 503: 1.

Kellerer, A.M. (2000). *Radiat Environ Biophys* 39: 17.

Kesavan, P.C. (1997). *High Levels of Natural Radiation* 1996 (Eds. L. Wei, T. Sugahara and Z. Tao) *Elsevier Science BV.* Amsterdam.

Kondo, S. (1993). *Health Effects of Low-level Radiation*, Kinki University Press, Osaka, Japan; Medical Physics Publishing, Madison, WI, USA.

Kodama, K., K. Ozara and T.J. Okubo (2012). *Radiol Prot* 32, N 51.

Kodaria, M., C. Satho, K. Hiyama and K. Toyama (1995). *Am J Hum Genet* 57: 1283.

L. Abramsson – Zetterberg, G. Zetterberg, S. Sundell–Bergman and J. Grawe (2000). *Int J Radiat Bio* I76: 971.

Little, M.P. and C.R. Muirhead (2000). *Int J Radiat Biol* 76: 939.

Liu, S.Z., Y.C. Zhang, Y. Mu, X. Su and J.X. Liu (1996). *Mutat Res* 358: 185.

L. Mullenders, M. Atkinson, H. Paretzke, L. Sabatier and S. Bouffler (2009). *Nature Reviews* 9: 596.

Lorenz, E., E. Miller, C.C. Condon and R. Schweisthal (1955). *J. Natl Cancer Inst* 15: 1049.

Lu, L., L. Lu, B. Hu, F. Yu and Y. Wang (2009). *Int J Radiat Biol* 85: 532.

Lobrich, M., N. Rief, M. Kuhne, M. Heckmann, J. Fleckenstein, C. Rube and M. Ude (2005). *Proc Natl Acad Sci* USA 102: 8984.

Luckey, T.D. (1992). *Raidation Hormesis,* CRC Press, Boca Raton, FL, USA.

Maas, C., E. de Vries, SWG Tait and J. Borst (2011). *Oncogene* 30: 3636.

Matsumoto, K., M.J. Ramsey, D.O. Nelson and J.D. Tucker (1998). *Radiat Res* 149: 602.

Mistry, K.B., A.R. Gopal-Ayengar and K.G. Bharathan (1965). *Health Phys* 11: 1457.

Mitchell, S.A., S.A. Marino, D.J. Brenner and E.J. Hale (2004). *Int J Radiat Biol* 80: 465.

Morgan, W.F. (2003). *Radiat Res* 159: 567.

————. (2003). *Radiat Res* 159: 581.

Mothersill, C and C. Seymour (2006). *Dose-Response* 4: 283.

Mothersill, C., C.B. Seymour and M.C. Joiner (2002). *Radiat Res* 157: 526.

Mori, E, A. Takahashi, N. Yamakawa, T. Kirita and T. Ohnishi (2009). *Radiat Res* 50: 37.

Nagasawa, H. and J.B. Little (1992). *Cancer Res* 52: 6394

Nair, M.K., K.S.V. Nambi, N.S. Amma, P. Gangadharan, P. Jayalakshmi, S. Jayadevan, V. Cherian and K.N. Raghuram (1999). *Radiat Res* 152 S 145.

Nayar, G.G. , K.P. George and A.R. Gopal–Ayengar (1970). *Radiat Botany* 10: 287.

Neumaier, T., J. Swenson, C. Pham, A. Polyzos, A.T. Lo , P. Yang, J. Dyball, A. Asaithamby , D.J. Chen, M.J. Bissel, S. Thalhammer and S.V. Costes (2012). *Proc Natl AcadSci* USA 109: 443.

Nogami, M, J.T Huang, S.J. James, J.M. Lubinski, L.T. Nakamura and T. Makinodan (1993). *Int J Radiat Biol* 63: 775.

Neel, J.V. (1998). *Proc Nat Acad Sci* 95: 5432. USA.

Neel, J.V., C. Satoh, K. Goriki, J. Asakawa, M. Fujita, N. Takahashi, T. Kageoka and R. Hazama (1988). *Am J Hum Genet* 42: 663.

Neel, J.V., W.J. Schull, A.A. Awa, C. Satoh, H. Kato, M. Otake and Y. Yoshimoto (1990). *AmJ Hum Genet* 46: 1053.

Nowosielska, E.M., A. Cheda, J. Wrembel – Wargocka, M.K. Janiak. (2011). *Int J Radiat Biol* 87: 202.

Otake, M., W.J. Schull and J.V. Neel (1990). *Radiat Res* 122: 1.

Pandey, R., B..S Shankar, D. Sharma and K.B. Sainis (2005). *Int J Radiat Biol* 81: 801.

Pierce, D.A., G.B. Sharp and K. Mabuchi (2003). *Radiat Res* 159: 511.

Planel, H., J.P. Soleilhavoup, R. Tixador, G. Richoilley, A. Conter, F. Croute, C. Croute, C. Caratero and Y. Gaubin (1987). *Health Phys* 52: 571.

Portess, D.I., G. Bauer, M.A. Hill, P.O. (2007). Neill *Cancer Res* 67: 1246.

Russell, W.L. and E.M. Kelly (1982). *Proc Natl Acad Sci* 79: 542. USA.

Radford, I.R. (2002). *Int J Radiat Biol* 78: 1081.

Redpath, J.L., Q. Lu, X. Lao, S. Molloi and E. Elmore (2003). *Int J Radiat Biol* 79: 235.

Redpath, J.L., D. Liang, T.H. Taylor, C. Christie and E. Elmore (2001). *Radiat Res* 156: 700.

Rothkamm, K. and M. Lobrich (2003). *Proc Natl Acad Sci* 100: 5057. USA.

Sauvaget, C., F. Kasagi and C.A, Waldren (2004). *Mutat Res* 551: 145.

Sasaki, M.S., Y. Ejima, A. Tachibana, T. Yamada , K. Ishizaki , T. Shimizu and T. Nomura (2002). *Mutat Res* 504: 101.

Sowa, M.B., W. Goetz, J.E. Baulch, D.N. Pyles, J. Dziegielewski, S. Youino, A.R. Snyder, S.M. de Toledo, E.I. Azzam and W.F. Morgan (2010). *Int J Radiat Biol* 86: 102.

Shankar, B., R. Pandey and K. Sainis (2006). *Int J Radiat Biol* 82: 537.

Szumiel, I. (2005). *Int J Radiat Biol* 81: 233.

Tanooka, H. (2011). *Int J Radiat Biol* 87: 645.

Thierens, H., A. Vral, M. Barbe, M. Meijlaers, A. Baeyens and L. De Ridder (2002). *Int J Radiat Biol* 78: 1117.

Tubiana, M., A. Aurengo, D. Averbeck, A. Bonnin, B. Leguen, R. Massee, R. Monier, A.J. Valleron and De Vathaire F. (2005). Academie der sciences and Academie Nationale de Medicine, Paris. p.94.

Wei, L. (1997). *High Levels of Natural Radiation* 1996, L. Wei, T. Sugahara and Z. Tao (eds.), *Elsevier Science BV*. Amsterdam.

Yamashita, J., Y. Takizawa, A. Yakisaka (1991). 34th Annual Meeting, Japan Radiation Research Society, Nov. 20-22, Tokyo.

Yoshimoto, Y, W.J. Schull, H. Kato and J.V. Neel (1991). *Mortality among the Offspring (F1) of Atomic Bomb Survivors, 1946-1985* (RERF TR 1–91) Radiation Effects Research Foundation, Hiroshima 1.

Zhang, L., Y. Tian, W. Yongyou, H. Zhang, Z. Wang, H. Huo, Y. Zhang, M. Zhang, P. Ning and J. Jiang (2010). *Int J Radiat Biol* 86: 329.

Zhou, H., G. Randers-Pehrson, C.A. Waldren, D. Vannais, E.J. Hall and T.K. Hei (2000). *Proc Natl Acad Sci* 97: 2099. USA.

*Acknowledgements:* The author dedicates this paper to Prof M.S. Swaminathan FRS, who introduced him to radiation biology five decades ago. The author also places on record his sincere thanks to Dr K.B. Sainis, Dr S.K. Apte of Bio-medical Group of Bhabha Atomic Research Centre, Mumbai and Prof K. Satyamoorthy of School of Life Sciences, Manipal University, Manipal, for helpful discussions.

6

S.D. GADEKAR AND S. GADEKAR

# Observations Regarding Health Impacts of Some Indian Nuclear Installations on Surrounding Populations

## Studies of Rawatbhata Nuclear Power Plant and Jaduguda Uranium Mines

## Background

The discovery of X-rays (Roentgen, 1895) and radioactivity (Becquerel, 1896) at the turn of the last century, vastly increased the impact of radiation on human health. The havoc this caused, especially on the health of pioneering radiologists, produced an intense interest in the question of health effects of radiation.[1] There were many lives lost in 'various incidents' before people started realising that radiation could cause serious health effects even after a long interval of time. These studies received a huge 'boost' after the atomic bomb explosions at Hiroshima and Nagasaki. Even today, radiation standards are determined mainly based on Hiroshima-Nagasaki studies (Richardson, 2012). However, these studies do not take into account effects of persistent low doses and slow dose rates. Presently, the radiation community accepts the linear no-threshold hypothesis although there are some who feel that radiation at low doses stimulates the immune system and results in beneficial effects (Hormesis) (Feinendegen, 2005; Kaiser, 2003). Despite strong advocacy with radiation standards setting bodies since early 80s, this hypothesis has still not been accepted

---

1. Clarence Dally, assistant to Thomas Alva Edison became the first person who died in 1904 at the age of 39 due to overexposure to X rays. He was followed by many others amongst radiologists.

by standards setting organisations. Recent epidemiological studies indicate that health effects may, in fact, be larger at low doses and slow dose rates than estimated by calculations based on Hiroshima Nagasaki data (Cardis *et al.*, 2005; Jacob *et al.*, 2009).

## Introduction

There have been essentially no efforts to study such effects in India. Although it is now almost four decades after the first mines and power plants started operating, there has still not been any systematic study. What has been done is with the high natural background that is observed on the beach sands in Kerala and on some other places on the Eastern coast (Kochupillai *et al.*, 1976; Nair *et al.*, 1999).

The reactors at Rawatbhata were the first power reactors of the CANDU type (Canadian Deuterium-Uranium) built in India. Since, the Indian nuclear power programme was based on the CANDU type of reactors, this type was the prototype for the whole programme. The site was selected in 1961 and construction on unit 1 started with Canadian help in 1964. The unit achieved criticality in August 1972 and was declared commercial in December 1973. Work on the second unit began in 1967 and it became commercial in April 1981. Besides the two reactors, the only other large industrial establishment in the area at the time our survey was conducted, (1991) was a heavy water plant to produce heavy water used as both moderator and coolant in the reactors.

India's nuclear programme derives its uranium from mines located in and near the town of Jaduguda in the state of Jharkhand. The Uranium Corporation of India Limited (UCIL) has been in charge of the mining and milling operations since commencement in the mid-60s. A number of news reports in various magazines and newspapers have detailed various health problems suffered by the people of the area (Chakrabarti, 1998; Rahman and Basu, 1999). These reports, have in particular, focused on congenital deformities in children. In December 1998, the state legislative council of Bihar, issued a document acknowledging the existence of health related problems (Hussain, 1998). Two films document some of the health related problems faced by people in the area and the poor management practices utilised to deal with the mined uranium and its by-products (Patwardhan, 2002; Shriprakash, 1999). Predictably, the UCIL management has denied such report (George, 1999).

Nuclear establishments the world over do a shoddy job of assessing the health and environmental damage caused by their activities. Even when some research is done, most establishments are loathe to publish whatever data they have so that it can be independently evaluated. Getting any information from the nuclear establishment is a Herculean task despite the Right to Information Act. (To give an instance, even information about emergency relocation plans [which supposedly is a public document] is not available to the general public. It is given only to local bureaucrats.)

For many years, the nuclear debate in India was a dialogue of the deaf, since neither side had any real data regarding the effects of nuclear activities on the environment or the health of the people living in the vicinity of nuclear facilities. As building of new plants was authorised, there were large protests in the vicinity of the proposed sites, but these protests lacked the punch that, a more informed debate based on observed facts could have provided. Our own involvement in anti-nuclear protest began in 1986, following the accident at Chernobyl. Our group, Anumukti, which is based in a small village in India called Vedchhi, near the then proposed nuclear power plant at Kakrapar, organised a protest rally near the plant site. More than ten thousand people came for the rally.

In September 1991, we conducted a survey of five villages near the nuclear power plant at Rawatbhata, situated 55 kilometres from the city of Kota in the state of Rajasthan.

## Methodology

The methodology we used in the study has been successfully utilised by many independent groups working in the field of maternal and child health. It involves systematically collecting information on a variety of indicators related to the health and socioeconomic status of people living in a certain region and comparing it with a control region chosen on the basis of justifiable criteria. The information was collected by a trained team of investigators, conversant in the local language, who ask selected individuals from these populations to answer a carefully designed detailed questionnaire.

The questionnaire that our study utilised consisted of over 170 questions concerning both household and individual level variables. At the household level, the aim was to determine the economic status of the household through collecting information on land holdings as well as various productive and non-productive possessions. There were also questions related to the availability of sources of drinking water and the type of fuel used for cooking. Individual level questions were related to age, place of birth, marital status, education, employment, relationships within the family, smoking and tobacco chewing habits, general health, duration of acute and chronic illness, and congenital and acquired deformities. Couples were also asked if they were consanguineous. There were separate questions relating to births and deaths in the household within the last two years before the survey. All of these questions were administered to each and every household in the area being studied, as well as in the controls.

All married women aged between 15 and 45 were asked questions related to their pregnancy history. Each and every pregnancy, along with the outcome of that pregnancy, was recorded. The dates of birth of children, who subsequently died were also registered. Women were also asked about their menstrual periods.

The same methodology was used in both these studies described below at Rawatbhata and at Jaduguda. At Jaduguda, besides the other questions, uranium workers were also asked separately regarding their terms of employment, the kind of work they did, for how long and whether they had regular medical and radiation checkups and also if they were informed of the results. The same was also needed to be done at Rawatbhata, but unfortunately was not done due to our inexperience.

The surveys also included a nutritional study of one-fifth of all households. The studied households were selected randomly. The questions were related to what the family had eaten, the day before. To check for consistency, we also asked them about their expenditure on various food items.

The methodology compares various calculated indices for both the proximate and distant areas at both sites. Ideally, the best case would be to compare the population of a certain area before a project begins and some years after the project has begun. However, such a study was never done before the start of mining and ore processing operations in the Jaduguda area or before the reactors were set up in Rawatbhata. This is part of a general lacuna with Indian developmental projects: base line health studies are not performed as part of environmental assessment. Even today, despite the widespread publicity about health awareness in Jaduguda, UCIL has not bothered to do baseline studies as a part of the process of setting up of new mines in the state of Meghalaya in North Eastern India or in Andhra Pradesh and Karnataka, where it is in the process of opening new mines or has already started mining. Similarly, there have been no baseline health impact studies at any proposed reactor sites.

In the absence of a base line study for comparison, the methodology requires choosing a control population that is similar in every respect to the population of the survey area, except with respect to the effects of the project (or other factor) being studied. This is difficult in general, and especially so in the vicinity of a large 'development' project because its impacts are widespread and deep.

## Rawatbhata

In September 1991, we surveyed a total of 1,023 households, of which 571 were in five villages within 10 kilometers of the Rawatbhata nuclear power and 472 were in four distant villages, which were more than 50 kilometers from the plant. The total number of people surveyed was 2,860 in the proximate villages *versus* 2,544 in the distant villages.

As is apparent from Tables 6.1 to 6.3, there was a broad similarity between the two areas with respect to age and sex distribution. There was no great difference in the areas with respect to caste distribution, either. Educational status was uniform in both areas; about 70 per cent were illiterate.

**Table 6.1**

*Age and Sex Distribution*

| Age Group | Near Villages | | Far Villages | |
|---|---|---|---|---|
| | *Males* | *Females* | *Males* | *Females* |
| < 1 year | 63 (2%) | 57 (2%) | 43 (2%) | 55 (2%) |
| 1 – 4 years | 136 (5%) | 134 (5%) | 126 (5%) | 122 (5%) |
| 5 – 14 years | 393 (14%) | 370 (13%) | 356 (14%) | 359 (14%) |
| 15 – 24 years | 284 (10%) | 283 (10%) | 216 (8%) | 222 (9%) |
| 25 – 34 years | 220 (8%) | 196 (7%) | 176 (7%) | 180 (7%) |
| 35 – 44 years | 141 (5%) | 122 (4%) | 140 (6%) | 108 (4%) |
| 45 – 54 years | 104 (4%) | 100 (3%) | 79 (3%) | 89 (3%) |
| Over 55 years | 150 (5%) | 107 (4%) | 145 (6%) | 128 (5%) |
| Total | 1491 (52%) | 1369 (48%) | 1281 (50%) | 1263 (50%) |

**Table 6.2**

*Caste Composition*

| Caste Group | Proximate Villages | | Distant Villages | |
|---|---|---|---|---|
| | *Males* | *Females* | *Males* | *Females* |
| Scheduled Tribes | 162 (6%) | 131 (5%) | 151 (6%) | 165 (6%) |
| Scheduled Castes | 421 (15%) | 371 (13%) | 268 (11%) | 263 (10%) |
| Upper Castes | 202 (7%) | 188 (7%) | 103 (4%) | 87 (3%) |
| Backward Castes | 703 (25%) | 674 (24%) | 755 (30%) | 745 (29%) |
| Muslims | 3 (0%) | 5 (0%) | 4 (0%) | 3 (0%) |
| Total | 1491 (52%) | 1369 (48%) | 1281 (50%) | 1263 (50%) |

**Table 6.3**

*Educational Status*

| Ed. Qualifications | Proximate Villages | | Distant Villages | |
|---|---|---|---|---|
| | *Males* | *Females* | *Males* | *Females* |
| Illiterate | 877 | 1214 | 681 | 1118 |
| Up to Class 5 | 368 | 118 | 412 | 128 |
| Up to Class 10 | 216 | 34 | 169 | 17 |
| Up to Class 12 | 24 | 3 | 15 | 0 |
| More than Class 12 | 6 | 0 | 4 | 0 |

Another point of striking similarity in both areas was the diet. (Table 6.4) We asked diet-related questions to a randomly selected 20 per cent of the households. While there was a large variation in the types and amounts of foods consumed within each area from house to house, the averages for both areas for protein, carbohydrate and fat intake amounts were identical and in close agreement with the average Indian dietary intake.

**Table 6.4**

*Diet*

| Category | Near Villages | Far Villages |
|---|---|---|
| Energy | 2474±991 Cal | 2405±784 Cal |
| Protien | 80±30 gms | 79±27 gms |
| Carbohydrates | 448±159 gms | 457±156 gms |
| Fats | 32±25 gms | 27±16 gms |
| Calcium | 549±346 mg | 522±275 mg |
| Iron | 38±35 mg | 39±37 mg |
| Vitamin A | 778±698 mg | 1257±1904 mg |
| Vitamin C | 13±20 mg | 24±34 mg |

The living conditions in both areas, too, were very similar and this can be seen by the size and type of houses, the time required to fetch drinking water, the fuel used for cooking, and other factors. (Table 6.5)

**Table 6.5**

*Housing Conditions and Water Supply*

*(Per cent)*

| | Proximate Villages | Distant Villages |
|---|---|---|
| *Housing Conditions* | | |
| One room | 61.2 | 56.1 |
| 2 to 4 rooms | 37 | 41.6 |
| More than 4 rooms | 1.6 | 2.3 |
| *Kutcha* Houses | 85.3 | 82.5 |
| Electricity Connection | 19.9 | 52 |
| *Water Supply* | | |
| Well | 59 | 71.3 |
| Hand-pump | 28.6 | 28.2 |
| Water tap | 9,8 | -- |
| Pond | 2 | -- |
| 5 minutes' walk | 66.5 | 52 |
| 15 minutes' walk | 21.4 | 40.8 |
| One hour walk | 10.9 | 7 |

Similarly, various maternal indices—like average number of pregnancies, average family size, age of women at marriage, mother's age at the birth of first child, mother's age at the time of miscarriage, and mother's age at the birth of still-born and deformed children—were very similar in both areas. (Table 6.6)

**Table 6.6**

*Some Family Indices*

|  | *Proximate Villages* | *Distant Villages* |
|---|---|---|
| Family Size | 5.19 ± 2.66 | 5.39 ± 2.67 |
| Women Head of Households | 5.4 % | 5.5% |
| Av. Number of Pregnancies | 5.3 ± 2.09 | 5.15 ± 2.16 |
| Women's Age at: | | |
| Effective Marriage (*Gauna*) | 15.1 ± 2.0 years | 15.4 ± 2.5 years |
| Birth of First Child | 18.4 ±3.1 years | 19.3 ± 3.2 years |
| Live Birth | 23.7 ± 5.7 years | 24.0 ± 5.5 Years |
| Abortion | 23.1 ± 5.9 years | 21.5 ± 5.2 years |
| Still-Birth | 22.8 ± 7.0 years | 22.0 ± 6.7 years |
| Birth of Deformed Child | 25.2 ± 5.8 years | 23.8 ± 4.9 years |
| Av. Number of Children Born to Women Aged 45-49 Years | 6.3 | 7.1 |

The landholding pattern in the two areas does show some differences. People living near the plant were more likely to own land, while in the far villages, there were more landless people. On the other hand, people living in distant villages were more likely to have irrigated land and use a greater amount of fertilisers and pesticides in their agricultural practices.

The only surprising difference in the two areas in terms of living conditions was in the state of electrification. In the villages, farther away from the electricity-producing nuclear power stations, 52 per cent of the houses were electrified, while those nearby had only 19 per cent houses electrified.

Employment patterns in both areas were of course different because of the presence of the nuclear power plant. While almost all the labourers in the distant area work in the village itself, 44 per cent of the labourers who live near the plant, work for the nuclear energy establishment. But amongst these labourers, there were a very few (just four) people with low level regular jobs. Most work as casual workers at the plant doing construction and cleaning. Eight per cent of these are casual workers.

The most immediately noticeable conspicuous difference was the striking contrast in the pattern of sickness and disease in the two areas. More people reported of illnesses in the area proximate to the Rawatbhata plant. While 25 per cent of the people in the distant area reported some illness, 45 per cent did so in the proximate area. There were 68 households out of 551 with at least one member reporting four different ailments, while the number of such households in the distant area was just nine out of 472. Table 6.7 provides a comparison of disease prevalence between proximate villages and distant villages.

**Table 6.7**

*Disease Pattern*

| Complaints | Proximate Villages | | Distant Villages | |
| --- | --- | --- | --- | --- |
| | Affected Persons | Average Age in Years | Affected Persons | Average Age |
| Short Duration Fever | 137 (4.8%) | 26 ± 19 | 117 (4.6%) | 26 ± 19 |
| Breathing Difficulties | 71 (2.5%) | 45 | 52 (2.0%) | 48 |
| Persistent Cough | 103 (3.6%) | 31 ± 19 | 60 (2.4%) | 42 ± 22 |
| Long Duration Fever | 120 (4.2%) | 25 ± 17.5 | 41 (1.6%) | 30 ± 17.5 |
| Pain in Joints | 116 (4.1%) | 43 ± 15 | 56 (2.2%) | 45 ± 16 |
| Digestive Problems | 360 (12.9%) | 29 ± 18 | 151 (6.0%) | 33 ± 19 |
| Weakness & Debility | 147 (5.1%) | 36 ± 17 | 96 (3.8%) | 46 ± 18 |
| Skin Diseases | 208 (7.3%) | 21 ± 19 | 75 (2.9%) | 21 ± 20 |
| Persistent Body Ache | 126(4.4%) | 34 ± 15 | 28 (0.9%) | 33 ± 15 |
| Solid Tumours | 30 (1.1%) | 41 ± 21 | 5 (0.2%) | 50 ± 18 |
| Chronic Eye Problems | 51 (1.8%) | 39 ± 21 | 20 (0.8%) | 42 ± 13 |
| Conjunctivitis | 16 (0.6%) | 15 ± 17 | 12 (0.5%) | 12 ± 12 |
| Cataract | 21 (0.7%) | 58 ± 15 | 8 (0.3%) | 68 ± 7 |
| Acquired Deformities | 31 (1.1%) | 41 ± 15 | 17 (0.7%) | 48 ± 18 |
| Polio | 24 (0.8%) | 21 ± 18 | 17 (0.7%) | 21 ± 15 |

Amongst the types of ailments reported, there was no difference in acute problems like short duration fevers, conjunctivitis, acute breathing difficulties, etc. However, there were large differences in chronic problems like long duration fevers, long lasting and frequently recurring skin problems, cataracts, continual digestive tract problems, pain in joints, body ache, and a persistent feeling of lethargy and general debility. The number of people reporting these chronic ailments was two to three times higher in the proximate area. Also, they were on average 10 years younger than those reporting similar disease in the distant area. The largest differences were seen in the case of solid tumours. In villages near the nuclear plant we saw 30 cases, one the size of football on the chest of a woman, and several tennis ball sized tumors, whereas in the control villages there were only five such cases and none so large. Acquired deformities, such as due to accidents or illnesses like small-pox were slightly higher in the proximate villages but this was mainly due to blindness caused by a smallpox epidemic in the past. The incidence of Polio was the same in both areas.

The most significant differences in health were related to untoward pregnancy outcomes. These were observed in the whole range, including significantly higher number of miscarriages, still-births, deaths amongst newborn babies and congenital deformities amongst both the living and those who had died within the last two years. For instance, the total number of congenital deformities was 50 amongst 45 children, who lived near the power plant and 14 deformities amongst 14 children who lived in the far off villages. These numbers are statistically significant, but the significance becomes stark when we consider the differences amongst those born after the start of the power plant. Although the number of deformities amongst those more than 18 years of age, were just 5 near the plant *versus* 4 away from the plant, the numbers were 39 *versus* 6 among those who were less than eleven years of age, when both units of the plant had started working (1981). Similarly, while 7 infants died within a day of birth near the plant during the two years previous to the 1991 survey, the number dying in the distant areas was just one. There were six stillbirths near the plant, compared to zero away from the plant, in the same two years interval. The chances of such differences occurring in two comparable populations purely by chance are less than one in a million. On the other hand, deaths of newborn infants who had survived for a week and then died (usually due to infection) were almost the same in both the areas (9 near the plant and 7 in the distant villages).

**Table 6.8**

*Untoward Pregnancy Outcomes Two Years Previous to Survey*

|  | Proximate Villages | Distant Villages |
|---|---|---|
| Number of Pregnancies | 285 | 292 |
| Number of Births | 258 | 197 |
| Number of Spontaneous Abortions | 27 | 5 |
| Number of Stillbirths | 6 | 0 |
| Number of Live Born Children who Died Later | 31 | 20 |
| Number of Stillborn Children with Deformities | 4 | 0 |
| Number of Liveborn Children with Deformities | 16 | 3 |
| Number of Deformed Children who Died Later | 9 | 1 |
| One Day Deaths | 7 | 1 |
| Deaths Among Children with No Deformities | 28 | 18 |

Pregnancy outcome data and infant death data are summarised in Table 6.8 and 6.9

**Table 6.9**

*Deaths Amongst Infants*

|  | Proximate Villages | Distant Villages |
|---|---|---|
| Stillbirths | 6 (1.9) | 0 (1.5) |
| One Day Deaths | 7 | 1 |
| Early Neonatal Deaths | 13 (8.2) | 5 (6.5) |
| Infant Mortality | 32 (28.2) | 19 (22.6) |

The number in the parenthesis are those expected on the basis of 1991 census data for Rajasthan and Madhya Pradesh.

**Table 6.10**

*Deformity Pattern*

|  | Proximate Villages | Distant Villages |
|---|---|---|
| Deformities in total population | 50 (44) | 14 (44) |
| Deformities in population above 18 years of age | 5 (5) | 4 (4) |
| Deformities in population below 18 years of age | 45 (39) | 10 (10) |
| Deformities in population below 11 years of age | 38 (33) | 6 (6) |
| *Deformity Pattern seen during the Last Two Years* | | |
| Live Born | | |
| With Deformity | 16 | 3 |
| Without Deformity | 236 | 194 |
| Stillborn | | |
| With Deformity | 4 | 0 |
| Without Deformity | 2 | 0 |

It is easy to see that the observed differences in the two populations' health status are not due to the 'usual suspects,' poverty, malnutrition or unsanitary living conditions despite government assertions to the contrary. In fact, because of the large injection of money due to the presence of the plant in the vicinity, the people near the plant were earning more than those living farther away, thus, whatever effects due to poverty should have been in the opposite direction to those observed. Unfortunately, even with higher earnings, they were financially not better-off in real terms (possessions) because of their high medical bills. Pesticide use was greater in the distant villages and any deformity due to that should have been more prevalent in the distant villages compared to the proximate villages.

**Table 6.11**

*Causes of Death in Children (Aged < 5 Years)*

| Causes of Death | Proximate Villages | Distant Villages |
| --- | --- | --- |
| Fevers | 3 | 7 |
| Diarrhoea | 6 | 3 |
| Tetanus | 1 | 6 |
| Respiratory Infection | 1 | 3 |
| Measles | 1 | 0 |
| Polio | 1 | 1 |
| Congenital Deformities | 10 | 1 |
| Small Baby | 10 | 1 |
| Unknown Causes | 3 | 1 |

**Table 6.12**

*Causes of Death amongst Adults*

| Causes of Death | Proximate Villages | Distant Villages |
| --- | --- | --- |
| Fevers | 11 | 9 |
| Diarrhoea | 5 | 6 |
| Respiratory Infection | 5 | 6 |
| Old Age | 8 | 14 |
| Pain Abdomen | 2 | 1 |
| Paralysis | 2 | 2 |
| Accidents | 1 | 2 |
| Perinatal Deaths | 2 | 0 |
| Cancers | 6 | 2 |
| Unknown Causes | 0 | 3 |

Again looking at the causes of death in the last two years, we find that amongst children, the most significant differences are deaths due to congenital deformities or due to the small size of the baby. Both these causes are interesting to note since they are related to assaults on the fetus in the womb, whereas, causes related to infections and unsanitary living conditions are the same in both areas. Tetanus is the exception. It was much higher in the distant villages.

An examination on causes of death amongst adults (Table 6.12) does not show anything significantly different but it is interesting to note that death due to 'old age' was higher in the distant villages. This is probably due to the fact that people die about ten years earlier in the proximate villages.

We published these results in *International Perspectives in Public Health 10* (1994) and there was a great deal of media interest in our findings. The government and the nuclear authorities at first stoutly denied that there were any health effects at all. Their argument was that if such effects were present, they would have known about them, themselves. After many independent newspapers and TV crews had been to the area (even *60 Minutes*), so that the fact of the health effects could no longer be denied, the authorities started saying that whatever was there had nothing to do with radiation. They preferred their old faithfuls: poverty, malnutrition and unsanitary living conditions. However, these assertions were made without any proper survey on their part.

The most effective use of these results occurred when we printed a summary of the results in Hindi and distributed it to each and every house in all the villages near the plant. Although most people were illiterate, they had the summary read in their presence. Since, they had been suffering the consequences, they could understand the conclusions of the survey only too well, but having scientific proof empowered them no end. Six months later they organised a rally by themselves where, for the first time ever, people demanded that the reactors be shut down. In this area, where there is a great deal of *purdah* (women not appearing in public), an old tribal woman spoke out at the public meeting, chastising the people for wanting electricity for which the price was, increased deformities in children.

## Jaduguda

*Criteria for Selection*

We chose villages as the unit of our survey and selected villages for study on the basis of the criteria mentioned below. All the households in the selected villages were studied. The choice of villages near Jaduguda was determined solely by the proximity of the village to the tailing ponds, where most of the material leftover after the extraction of uranium from the ore is dumped. Since, we considered the whole village as a unit, all the villages with at least some portion near the tailing ponds were selected. In case of spatially spread out villages, this procedure meant that we have selected households in our survey, which do live farther away from the tailing ponds than some other households in villages not part of the survey. The villages chosen near Jaduguda ('nearby' from here on) were:

1. Mechua.

2. Bhatin.

3. Tilaitand.

4. Rohinibeda.

As control cases ('controls' from here on) we chose two villages:

1. Chandpur.

2. Purana Chaibasa.

These control villages match the villages of the survey area well if one looks at age and sex distribution of the population and other demographic indicators (see Table 6.13). The same is the case with the caste composition (see Table 6.14).

**Table 6.13**

*Age and Sex Distribution of the Population and Other Demographic Indicators*

| | Nearby Villages | | Control Villages | |
| Ages | Males | Females | Males | Females |
| --- | --- | --- | --- | --- |
| 0-6 | 421 (16.2%) | 407 (16.0%) | 209 (16.3%) | 224 (18.6%) |
| 7-14 | 488 (18.8%) | 512 (20.1%) | 276 (21.5%) | 211 (17.6%) |
| 15-49 | 1346 (51.8%) | 1238 (48.6%) | 626 (48.8%) | 564 (46.9%) |
| 50-59 | 216 (8.3%) | 231 (9.1%) | 103 (8.0%) | 123 (10.2%) |
| >=60 | 125 (4.8%) | 157 (6.2%) | 68 (5.3%) | 80 (6.7%) |

**Table 6.14**

*Caste Composition*

(*Per cent*)

| Caste Group | Nearby Villages | Control Villages |
| --- | --- | --- |
| Scheduled Tribes | 75.7 | 77.0 |
| Scheduled Castes | 3.4 | 4.7 |
| Backward Castes | 13.2 | 12.7 |
| Upper Castes | 3.7 | 4.7 |
| Others[1] | 3.9 | 0.9 |

*Note*: 1. Others are mostly people who have moved from elsewhere and do not belong to any local caste. Their numbers are naturally more in nearby villages since they are likely to have come for jobs and the ancillary employment generated by large development projects.

The big difference between the control villages and the villages near the tailing ponds is in the level of economic development. It can be stated unambiguously that uranium mining has brought prosperity to Jaduguda.

Despite Purana Chaibasa being just 3 km from Chaibasa town (which incidentally is a much larger town than Jaduguda), the population of the control villages, depend more on agriculture. 44.5 per cent of the head of households work in agriculture or related activities such as animal husbandry. The percentage of such people near Jaduguda is only 27.5 per cent. On the other hand, the per cent age of head of households having a regular salaried job is 29.1 per cent near Jaduguda, but only 15.1 per cent in the controls. Possession of productive items required for a rural agricultural economy like ploughs, bullocks, cows, goats, agricultural and non-agricultural implements, etc., are slightly higher in the control villages, though the difference is not statistically significant. There is a sharp and marked difference in the ownership of non-productive consumer items such as clocks, radios, TV sets, two-wheelers, electric appliances, gas stoves etc. At least 50 per cent or more, of the households near Jaduguda possess such items, and more of them as compared to households in the control areas. For instance, 207 households near Jaduguda possess TV sets whereas this number is only 52 amongst the controls. Since, we have surveyed twice, as many households nearby, one would expect around a hundred households with TV sets nearby, but the 207 households with TV sets show that the area is definitely more prosperous. This divergence becomes even more pronounced if one looks at higher value items such as refrigerators or two wheeled motorised vehicles. Whereas, there are 12 households having refrigerators nearby, there is only one such house in the controls. The differences in the two areas for basic electricity services like lighting, are not so pronounced. Whereas, 42.7 per cent of the households nearby are electrified, their number in the controls is 38.6 per cent. Urbanised modes of lifestyle such as having bathrooms (50 *versus* 6) or latrines (32 *versus* 6) are also higher in nearby villages.

The main reason for the increased prosperity is simple. Unlike the policies practiced at other nuclear facilities like Rawatbhata in India, where the local inhabitants get almost no regular jobs, 55.3 per cent of the households in the nearby villages have at least one member employed with regular salaries by the Uranium Corporation of India Ltd (UCIL), either as miners or as mill workers. This means that the household earns at least ₹ 6000 per month (the minimum starting salary at the time of the survey). Most earn more. Also, regular employment guarantees other benefits such as access to company medical facilities. Of the 920 households in all, there are 83 households with more than one person from the house working with UCIL.

**Implications for Analysis**

Although our controls match the survey area quite well in terms of caste and tribal composition of the population as well as the age and sex distribution of the population, to get an economically matched tribal population, is extremely difficult, since no other nearby group has 'benefited' to a similar extent from any project.

The sharp differences in economic well-being are a problem as far analysis is concerned. The very essence of a control population is that it mimics the survey population in all other respects, but, for the effect of the agent being studied. That is just not possible in this case, since, there is no other area having similar tribal mix with the same level of prosperity in a generally rural setting like Jaduguda.

Other indicators of economic well-being and social development are also better in the nearby villages. We did a random sampling of 20 per cent of the households for what they had eaten the previous day. Although this single day estimation is not a fool proof method of determining the nutritional status, the average calorific intake is higher in the nearby villages when compared to that in the control villages (2492 calories per day per capita as against 2248). It may be mentioned that the calorific intake of the nearby villages are higher than the national average of 2400 calories per day. The number of illiterate women is lower in the nearby villages (59.3%) as compared to the control villages (67%). Numerous studies have shown that higher levels of women's education result in better hygiene and health status in the household.

**Results**

*Deformities*

Since, the bulk of the anecdotal evidence presented in magazines and newspapers and the documentary films mentioned earlier concerned high rates of congenital deformities, estimating the rates of such deformities was the primary focus of our survey. The numbers of persons with congenital deformities in the two areas surveyed, are listed in the table below. The number of houses surveyed near Jaduguda was double that of houses surveyed in the controls. This is also reflected in the population, which just shows that the household size in both the areas is very similar. Thus, one expects the number of deformities to be twice in nearby villages as compared to the controls. However, there were 63 people with deformities in nearby villages, whereas there were 13 such people in the controls; the ratio is much greater than the expected two. The probability of this happening just by chance is less than 1 in 700.

**Table 6.15**

*Numbers of Persons with Congenital Deformities in the Two Areas Surveyed*

| | Nearby Villages | | Control Villages | |
|---|---|---|---|---|
| Total Number of Houses Surveyed | 920 | | 469 | |
| | *Males* | *Females* | *Males* | *Females* |
| Total population | 2602 | 2548 | 1286 | 1206 |
| Population older than 35 years | 594 | 624 | 289 | 310 |
| Population 35 years or younger | 2008 | 1924 | 997 | 896 |
| Persons with congenital deformities | 35 | 28 | 7 | 6 |
| Persons over 35 with congenital deformities | 2 | 1 | 1 | 2 |
| Persons younger than 35 with congenital deformities | 33 | 27 | 6 | 4 |
| Persons with multiple congenital deformities | 13 | 7 | 0 | 1 |
| Total number of deformities | 52 | 34 | 7 | 7 |

The data collection for the survey was done in September 2000, whereas, mining in Jaduguda started in mid-60s. People older than 35 years of age, were born before the commencement of mining. Amongst them we see that there are only three people each, in both the areas with deformities. This may seem surprising since, we expect to see twice the number near Jaduguda. However, the numbers involved are small and even the difference of one case this way or that can change the whole picture.

But, when we look at people born after the start of mining operations, we find that there is a statistically significant rise in congenitally deformed amongst them. If we take the multiple deformities into account, then, the probability of such a huge rise in deformities having occurred just by chance is about one in a hundred thousand. One can be fairly certain that the increased level of deformities in children is a real effect and not some statistical anomaly.

Additional support for this is gained by looking at Table 6.15. The table shows that deformed children are still being born in Jaduguda and its environs. The huge increase in deformities is not due to some past accident, effect of which will slowly vanish. It is a continuing disaster.

**Table 6.16**

*Pregnancies and Pregnancy Outcomes in the Two Years Preceding Survey*

|  | Proximate Villages | Distant Villages |
|---|---|---|
| Number of pregnancies | 265 | 144 |
| Number of births | 266 | 144 |
| Number of stillbirths | 4 | 2 |
| Number of live born children who died later | 9 | 6 |
| Number of stillborn children with deformities | 1 | 0 |
| Number of live born children with deformities | 14 | 1 |
| Number of deformed children who died | 8 | 1 |
| One day deaths | 6 | 2 |
| Deaths among children with no deformities | 1 | 5 |

There are also differences in the numbers of other untoward pregnancy outcomes. In the last two years, before the survey was done, there were a total of 409 pregnancies. Of these, there were 265 pregnancies including a pair of twins near Jaduguda and 144 in the control area. Thus, the birth rate is lower near Jaduguda than in the controls. There were four cases of stillbirths (one with deformity) near Jaduguda as compared to two in the controls. There were 15 cases of congenital deformities amongst 266 children born near Jaduguda, while there were just two cases amongst 144 born in the control area. Of the 14 live births with deformity in the nearby villages, eight died and only six survived. One of the children with deformity in the control area also survived. The total number of live births in the nearby villages were therefore 262, while this number in the controls was 142. Of these 262, nine had died within a year whereas six had died in the control area. Of the nine who died near Jaduguda, six died within a few hours of birth, one survived a month, and another survived for nine months. Both of the survivors had deformities. One child survived for a year before death (of diarrhoea) and had no deformities. In the control villages, two were premature babies, who died within a few hours after birth; three others had died a week after birth out of diarrhoea and one had survived for a month.

There are two major causes of death of infants within a few hours of birth: obstetric problems during delivery or a defective foetus. The chances of obstetric problems during delivery being the cause of the higher number of one day deaths in Jaduguda are less because being prosperous it has attracted better medical care. A lot of household heads are company employees and are entitled to and avail of company medical facilities. Patients have also been sent to well-equipped hospitals in the city of Jamshedpur (40 km away) in an emergency.

An examination of the causes of death amongst children shows wide divergence in the two areas. In the nearby villages, of the nine children who died within a year of birth, eight had congenital deformities. On the other hand, in the controls of the six deaths, only one can be attributed to deformity, whereas five were due to environmental reasons such as diarrhoea, fever and premature birth.

A standard criticism from members of the nuclear establishment of other such studies have been that deformities amongst children have more to do with consanguineous parents and children being born rather late to mothers who have already crossed forty years of age. In the survey special efforts were made to check both these assertions. We found just one case of consanguineous marriage resulting in two children with squints. Similarly, there was just one case of a woman with two children, having squints, who was 38 at the birth of one of the children and 41 at the birth of the other. All these children were in the nearby villages. Even if we exclude these cases, it would make no difference to our conclusions that living near Jaduguda increases your risk of having a child with deformity.

Table 6.17 gives a tabulation of age of mothers in the two areas to show that whatever effects are seen they have nothing to do with variation in age.

**Table 6.17**

*Maternal Indices*

| Maternal Indices | Nearby Villages | Control Villages |
| --- | --- | --- |
| Average age at marriage | 16.3±2.9 | 16.8±2.7 |
| Average age of mother at birth of first child | 18.4±2.7 | 18.7±2.9 |
| Average age of mother at birth of last child | 24.8±5 | 25.1±5.2 |
| Average age of mother for all births | 22.6±4.9 | 22.9±4.9 |
| Average age of mother at birth of deformed child | 22.4±4.8 | 23.2±5.8 |
| Amongst surviving deformities average age of mother at their birth | 25±6.6 | 26±6.7 |

## Chronic Lung Disease

Studies all over the world of uranium miners have shown that they have higher incidence of diseases like silicosis and lung cancer. This fact has been known for almost two hundred years, since uranium mining started in Czechoslovakia. Even when lung cancer or silicosis was relatively unknown, people knew that uranium miners were prone to 'mountain sickness', which cruelly cut short their lives. A finding of high incidence of chronic lung problems amongst uranium miners would not be a surprise. They work in an environment, which is full of dust caused by blasting of rock, and in an area with elevated levels of radon.

What is then, indeed, rather surprising is that not a single case of any such work related disease has been diagnosed amongst uranium miners and mill workers in any of the people we examined near Jaduguda. Since, they are regular employees of UCIL, they are entitled to, and do avail off the company medical facilities, where the diagnosis is done by UCIL doctors for them and their families. Since 55 per cent of the households in the nearby villages have at least one worker working for UCIL, this means that a majority of the households do fall under these doctors's care. What we see instead, amongst miners and mill workers is an anomalously high incidence of tuberculosis.

*A word regarding methodology:* The questions asked in the questionnaire were: Is or was anybody in the family ill? For how long? What kind of treatment were given? Are any medical records available? Is the person currently ill and taking treatment? This methodology leads to a somewhat higher estimation of number of ill patients, since both those who are currently ill and those who have been cured of TB in the past, do get counted. National Family Health Survey, which follows a similar methodology, and who had conducted data collection at around the same time as us, found that there were 1,036 cases of tb per 100,000 in Bihar. At the time of the survey, Jharkhand had not been separated from Bihar.

The total population surveyed near Jaduguda was 5,150, while that in the controls was just a little less than half of that 2,492. In the controls we found 23 cases of TB (923/100,000). Near Jaduguda, we found 84 people (1,631/100,000). This was already on the higher side; not all of these cases were currently ill. Fifty-seven cases near Jaduguda and 15 cases in the controls said that they were currently suffering and taking treatment for the disease. Of these, 44 cases near Jaduguda and just six cases in the controls said that not only were they currently ill, but that they had been taking treatment for two years or more but without any improvement.

As is well known, TB is a very persistent disease. Patients immediately feel better on taking prescribed drugs and a good number of them are prone to discontinue treatment at some intermediate stage, which leads to recurrence of the symptoms of the disease, sometime later.

Let us divide the 5150 people living near Jaduguda into two groups: uranium workers who come in contact with uranium directly (miners, mill workers, casual labourers who load and unload drums of yellow cake on to trains for transport to the Nuclear Fuel Complex at Hyderabad, or the truck drivers who bring uranium ore from the mines at Narwapahar and Bhatin to Jaduguda and take the tailings slurry back for backfilling the mine), and others who do not come into direct contact with uranium. The latter group includes the family members of the first group along with others such as shop keepers and farmers who for some reason or the other have never caught the uranium frenzy. The number of uranium workers is 603, whereas the other group has 4,547 people in it. Of the total of 84 who complained of 'TB' most of whom—thanks to the good coverage provided by the employee's insurance scheme—have medical records to prove their claims; 49 belong to the first group of workers and 35 to the second group of non-workers. These numbers are summarised in Table 6.18.

**Table 6.18**

*Uranium Workers*

| | *Uranium Workers* | *Uranium Workers with TB* | *Uranium Non-Workers* | *Uranium Non-Workers with TB* |
|---|---|---|---|---|
| Nearby Villages | 603 | 49 (8126/100,000) | 4547 | 35 (770/100,000) |
| Control Villages | 13 | 0(0/100,000) | 2479 | 23 (923/100,000) |

The numbers in the parenthesis are incidence per thousand. As we can see, the incidence of TB amongst uranium non-workers is lower than that amongst the population in the distant villages. This is to be expected, since, there is higher level of prosperity in the proximate villages. They eat better, and have roomier houses. Unfortunately, the price for their prosperity is paid for by the greater than ten times higher rate amongst the bread-winners. The chances of this anomalously high rate amongst uranium workers being just an inadvertent statistical construct are less than one in a hundred million.

Doctors in our team made efforts to check the people who claimed they had TB. They found that some of the cases had already been cured. The number of cases who were identified by doctors to be currently ill with persistent lung problems were 25 (4146/100,000) amongst uranium workers, 11 (242/100,000) amongst non-uranium workers living in nearby villages and 8 (323/100,000) amongst people living in control villages.

Directly Observed Therapy (DOTS) programme for tuberculosis eradication in India gives a national average of 176/100,000. The data for Jharkhand is not separately available but from all accounts Jharkhand is a TB endemic area. Our data for the controls and for non-uranium workers in nearby villages are well within the range found by this programme.

The simplest explanation for the large number of 'TB' cases amongst uranium workers is that not all of them are TB, but include those mysteriously missing occupational lung diseases like silicosis and lung cancer that have been seen amongst uranium miners all over the world. The blame for this mis-diagnosis can be laid squarely on the shoulders of the UCIL doctors.

The consequences of this misdiagnosis are not benign. Being diagnosed with TB, patients continue to take TB treatment for long periods with no relief. This is also reflected in the deaths during the two years, preceding the survey.

A total number of 148 persons died in both the areas combined: 98 in the nearby villages and 50 in the controls. This is to be expected since we surveyed twice as many houses near Jaduguda. But, a detailed look at the pattern of death is very revealing.

Let us divide the group into two: children who have died before the age of twenty and adults who have died after reaching that age. Amongst children we find that 20 died near Jaduguda as compared to 13 in the controls. Out of these 20, 8 or 40 per cent died due to deformity (most during the first few hours of life itself) whereas amongst the 13 who died in the controls, only one died due to deformity. We have written about this earlier, but one needs to look at just one aspect—the male-female ratio. Of the 13 who died in the controls, there were six boys and seven girls, but of the twenty who died nearby Jaduguda, there were seven girls but 13 boys. The preferential difference for boys over girls is accounted for by looking at those who died due to deformity. Amongst those eight are seven boys and just one girl. Several other deformity studies from other parts of the world have shown that boys are more likely to be born with deformity than girls and the chances of survival of these babies is any way slim. The 'weaker' sex is stronger, where it counts.

Amongst adults, the total number of deaths in both areas is 115. Of these 78 are in the nearby villages, while 37 have occurred in the controls. Out of the 78 near Jaduguda, 53 were males and only 25 were females, while in the controls there were 22 males to 15 females. Why does death prefer men over women so strongly near Jaduguda?

The answer to this riddle is provided by looking into deaths due to 'TB', or to give it the correct name chronic lung problems. In the controls, the number of such deaths amongst men are 4 out of 22 (18.2%), while it is 2 out of 15 (13.3%) in women. Near Jaduguda death due to lung diseases amongst women is comparable 3 out of 25 (12%), but amongst men it is a whopping 22 out of 53 (41.5%).

TB today is a curable disease. Reluctance of some patients to continue the long treatment does result in persistent recurrence and ultimately drug resistance and death. But, the numbers here are too stark and show something quite abnormal. Fifty years ago, Swedish doctors did make mistakes and diagnosed these chronic lung problems as tuberculosis. But, this was recognised as an error and the diagnosis subsequently changed to silicosis. Dead men do tell tales!

No discussion of radiation effects can be complete without a mention of cancer. The radiation community considers an increase in cancer as the only effect of radiation exposure. Mr S.K. Malhotra, Head, Publicity Division, Department of Atomic Energy, writes in a letter to the editor of *Hindu* that, "It may be worthwhile to note that while the Indian Council of Medical Research  (ICMR) has estimated the national average incidence of cancer to be 74 per one lakh population, in Jaduguda the incidence is only 22."

It is unclear from this quotation, whether ICMR has estimated only the first portion (the national average of 74 per one lakh population) or also the second part regarding Jaduguda. It is likely to be just the former. During the survey, we found that there were six deaths due to cancer in the previous two years, five of them near Jaduguda and one in the controls. Amongst the living, there were six reported cancer cases, all of them near Jaduguda. However, when our team of doctors visited the houses, they could confirm only four cases. All of them had medical records from the Tata Main Hospital in Jamshedpur. The other two were not available at the time of the visit. Even if we discount these and take only the confirmed four cases it, works out to an average of 80 per one lakh population—a little higher than the ICMR's national average. Mr Malhotra's claim of 22 per one lakh population turns out on examination to be erroneous.

## Conclusion

The surveys have unambiguously demonstrated that there is a higher rate of congenital deformities among the people living in the vicinity of both, the uranium mines at Jaduguda and the nuclear power plant at Rawatbhata. It has also demonstrated a much higher rate of lung related problems among miners that have been misdiagnosed as TB. Both of these are statistically significant effects. Besides, it has shown that chronic diseases are more prevalent and appear at a younger age near nuclear facilities. Recently, an NGO in Chennai called ASPIRE under the leadership of Dr Manjula Dutta has done a survey of villages near Kalpakkam.[2] They too find a similar disease pattern as described above for Rawatbhata.

## References

Andrew, G. (1994). Centers for Disease Control and Prevention (CDC), Atlanta.

Archer, V.E. *et al.* (1998). *Journal of Occupational & Environmental Medicine* 40: 460-474.

Becquerel, H. (1896). "Sur les radiations émises par phosphorescence", *Comptes Rendus* 122: 420–421.

Cardis, E. *et al.* (2005). "Risk of Cancer after Low Doses of Ionising Radiation: Retrospective Cohort Study in 15 countries", BMJ 2005; 331 *doi: 10.1136/bmj.38499.599861.E0 (2005)*

Chakrabarti, A. (1998). *Indian Express*.

Directorate General of Mines Safety, M.o.L., Government of India. (1987).

Dye, C. *et al.* (1999). *JAMA* 282: 677-686.

Feinendegen, L.E. (2005). *British Journal of Radiology* 78(925): 3–7 *doi:10.1259/bjr/63353075.* PMID 15673519.

George, N. (1999). *Indian Express.*

Ghosh, D.K. (1999). *BARC Newsletter.*

Hnizdo, E. and G Sluis-Cremer (1993). *Am J Ind Med* 24: 447-457.

Hussain, J. (ed. Committee, B. L. C. E.) (Publications Division, Bihar Legislative Council, 1998)

Jacob, P., W. Ruhm, L. Walsh, et al. (2009). *Occupational & Environmental Medicine* 66(12): 789-796.

Jain, A.K. and M. Nag. Center for Policy Studies New York, The Population Council, New York.

Kaiser, J. (2003). *Science* 302(5644): 376–379 *doi:10.1126/science.302.5644.376. PMID 14563981.*

Kochupillai, N., I.C. Verma, M.S. Grewal and V. Ramalingaswami (1976). *Nature* 262: 60-61 doi:10.1038/262060a0.

Kretchik, J. (2003). *OSHA Chemical Health & Safety* 10.

Kumar, R. and A.H. Khan (1999). *24th Conference of the Indian Association for Radiological Protection.*

---

2. A Comprehensive Survey of Villages around Kalpakkam. Provisional Report (2012).

Lary, J.M. and L.J. Paulozzi (2001). Sex differences in the prevalence of human birth defects: A population-based study *Teratology* 64: 237–251.

Murthi, M., A.C. Guio and J. Dreze (1995). *Population and Development Review* 21: 745-782.

Nair, K *et al.* (1999). *Radiation Research* 152(6): S145-8.

Office of the Registrar General India. (2002).

Patwardhan, A. (2002). India.

Rahman, A.U. and J. Basu (1999). *Sunday.*

Richardson, D. (2012). *Bull. At. Sci.* 68(3): 29-35.

Roentgen, W. (1895). "On a New Kind of a Ray" Proceedings of Wurtzburg Physical-Medical Society.

Sciences, I. I. f. P. and O Macro (1999).

Shriprakash (1999). Krittika films, India.

Steenland, N. and D. Brown (1995). *American Journal of Public Health* 85: 1372-1377.

# III

# Fuel Cycles and Technology

7

FRANK VON HIPPEL

# History and Current Status of Reprocessing and Plutonium Breeder-Reactor Programmes Worldwide

## Introduction

### Why Civilian Reprocessing Began in the 1970s

Reprocessing was originally developed during World War II to obtain plutonium for the US nuclear-weapon programme. But even during World War II, some of the scientists involved hoped that, after the war, the focus would shift to using uranium as fuel for power reactors. It was believed that uranium was too scarce to support large-scale nuclear energy based primarily on the fission energy obtainable from the rare chain-reacting isotope U-235 (0.7% of natural uranium). In 1944, therefore, Leo Szilard (1972) invented the liquid-sodium-cooled plutonium breeder-reactor to exploit the fission energy latent in non-chain-reacting U-238 that makes up 99.3 per cent of natural uranium.

Belgium, France, Germany, India, Italy, Japan, Russia, the UK and the US, separately and in various combinations, all launched breeder-reactor and civilian reprocessing programmes to recover the plutonium in spent power reactor fuel in the expectation that plutonium would be needed for initial cores of the plutonium breeder-reactors, which were expected to be commercialised in the 1970s.[1]

---

1. Richard Garwin pointed out at the time, however, that it would be more uranium efficient to enrich uranium to 20 per cent to provide the startup cores for breeder reactors than to use the uranium in burner reactors to produce plutonium for that purpose, "The Role of the Breeder Reactor" in F. Barnaby *et al.* (1979). *Nuclear Energy and Nuclear Weapon Proliferation*. Taylor and Francis. p.141.

## Why Breeder Reactors have Failed Commercially

In 1975, the IAEA projected that global nuclear power capacity would grow to 1600 GWe by 2000; while also estimating that the global uranium resource amounted to 3.5 million tonnes, only enough to support 500 GWe of light-water reactor capacity for about 40 years (Fitts and Fujii, 1975). But, nuclear power did not grow that rapidly and more uranium was found. Today, global capacity is 375 GWe, the International Atomic Energy Agency (IAEA) projects that the capacity in 2050 will be between 560 and 1230 GWe (IAEA, 2011), and global uranium resources recoverable at current low prices are estimated at 17.5 million tonnes, enough to support 2500 GWe for another 40 years (Figure 7.1) (Nuclear Energy Agency and IAEA, 2009).

**Figure 7.1**

*IAEA Projections for Global Nuclear Capacity and of World Uranium Resources Translated into the Light-Water Reactor Capacity that could be Fueled for 40 Years on a Once-Through Fuel Cycle*

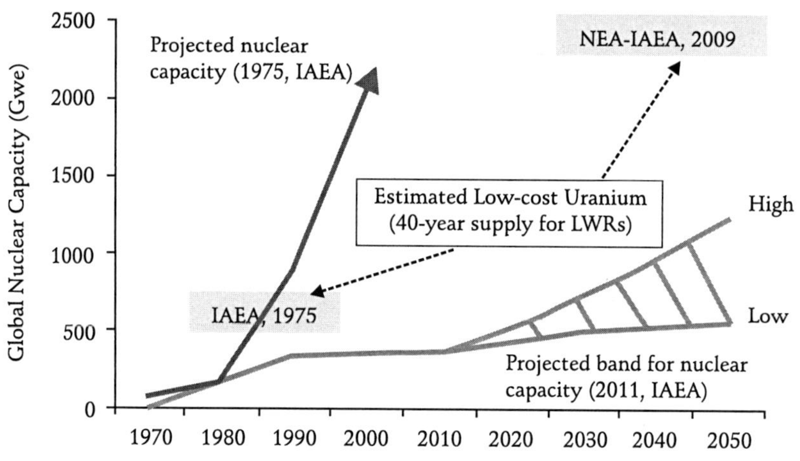


*Source*: Author

The price of natural uranium has fluctuated. The underlying trend has not been upward, however (Figure 7.2). At a price of $130/kg, uranium contributes about $0.003 to the price of a kilowatt-hour of nuclear electricity. The elimination of this cost for breeder reactors would be approximately offset by the cost of reprocessing breeder reactor fuel. In order to compete with a water-cooled reactor, therefore, the capital cost of a sodium-cooled reactor would have to be approximately the same.

There is, in fact, one important factor that would tend to reduce the relative capital cost of a sodium-cooled reactor: it operates far below the boiling temperature of sodium (883 °C) and therefore, at low pressure, while water-cooled reactors operate above the boiling temperature of water and hence, require thick pressure vessels and piping.

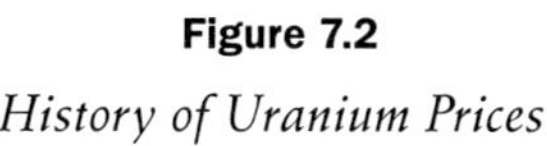

**Figure 7.2**

*History of Uranium Prices*

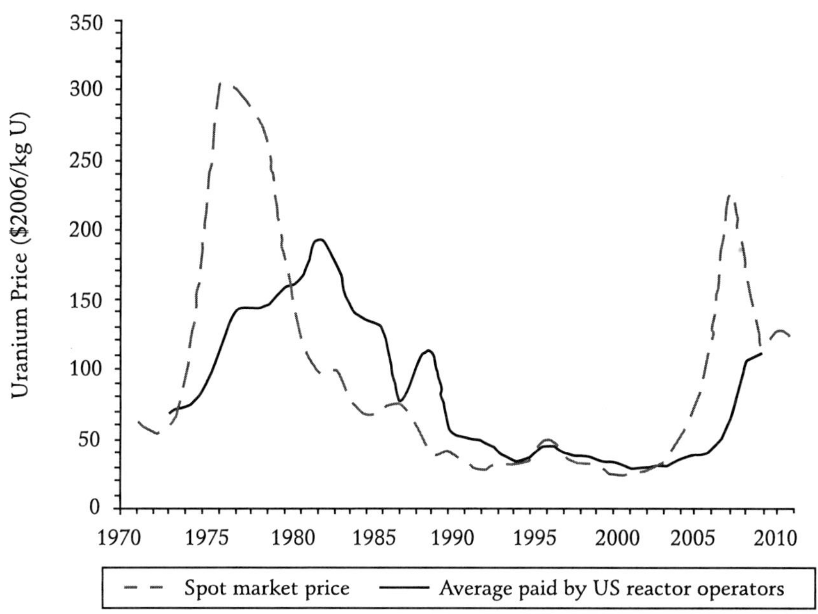


*Note:* The peak in the 1970s was in response to the projections of about 2,000 GWe of nuclear capacity by the year 2000 (Figure 7.1). The more recent peak was associated with the predictions of a nuclear 'renaissance' and the fact that many mines had been closed down as a result of the selling off of excess stocks acquired in the 1970s and then the sale of blended-down Soviet weapons uranium.[2]

However, another key factor cuts in the opposite direction: sodium burns when exposed to air or water. Leakage is therefore, much more damaging than from a water-cooled reactor. Also, it is

---

2. See update of figure 5 in M. Bunn, S. Fetter, J.P. Holdren and B. Van Der Zwaan, *Nuclear Technology* 150 (2005) p.209, with permission from S Fetter

impossible to simply open up the reactor during refueling or for maintenance, as is done with light-water reactors. As a result, as Admiral Hyman Rickover stated in 1956, after trying a sodium-cooled propulsion reactor in the second U.S. nuclear submarine, sodium-cooled reactors are 'expensive to build, complex to operate, susceptible to prolonged shutdown as a result of even minor malfunctions, and difficult and time-consuming to repair' (Hewlett and Duncan, 1974).

**Table 7.1**

*Generating Capacities, Years of Operation and Average Capacity Factors of the Highest-Power Prototype and Demonstration Breeder Reactors Built by Each of Six Countries*[3]

| Country | Demonstration Breeder Reactor | Power (MWe) | Years of Operation | Capacity Factor (%) |
|---|---|---|---|---|
| France | Superphenix | 1200 | 1985-1998 | 8 |
| Germany | SNR-300 | 300 | (1991) | 0 |
| Japan | Monju | 250 | 1994-95 | 1 |
| Russia | BN-600 | 600 | 1980- | 72 |
| UK | PFR | 500 | 1975-1994 | 19 |
| USA | Fermi I | 66 | 1966-1972 | 1 |

Experience has mostly borne out Rickover's assessment. Of the highest-power sodium-cooled reactors built by France, Germany, Japan, Russia, the UK and the US, only one achieved any longevity or a capacity factor above 20 per cent. That reactor is Russia's BN-600 reactor, which has operated for more than 30 years with a cumulative capacity factor, slightly greater than 70 per cent despite 14 sodium fires in its first 17 years.

The Organisation for Economic Cooperation and Development (OECD) countries, including France, Japan, the UK and the United States, on an average devoted about half of their nuclear-energy research, development and demonstration (RD&D) budgets to breeders until the early 1980s, but, today their expenditures on breeder RD&D are almost zero (Figure 7.3). India and Russia, which are not members of the OECD, however, are building large

---

3. Information taken from the IAEA's Power Reactor Information System

prototype reactors (500 and 800 MWe, respectively). Overall about $100 billion has been spent on breeder reactor R&D thus far.[4]

**Figure 7.3**

*Total Nuclear Fission and Breeder R&D in OECD Countries since 1974*[5]

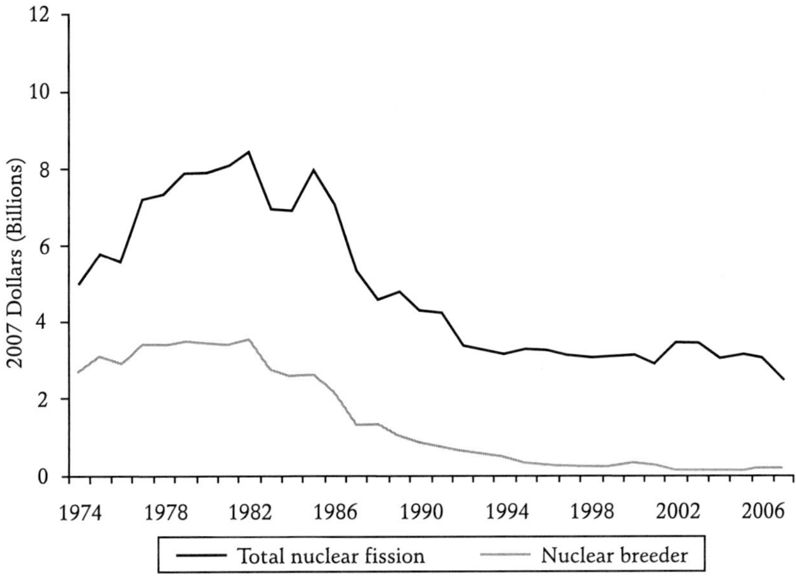


## *Country Choices with Regard to Reprocessing*

With the failure of their breeder commercialisation efforts, France and Japan decided nevertheless to continue to reprocess and to recycle the resulting separated plutonium into fresh fuel for their light-water-cooled reactors (Figure 7.4). France has begun to re-enrich the uranium it recovers by reprocessing and recycling of the product as well. This reduces the uranium requirements of a light water reactor by about 25 per cent but that saving is more than offset by the cost of reprocessing.

---

4. Fast Breeder Reactor Programmes: History and Status (2010), pp 6-7.

5. Fast Breeder Reactor Programmes: History and Status (2010), Figure 1.3. Based on International Energy Agency R&D Statistics Database.

**Figure 7.4**

*Storage of Spent Fuel*

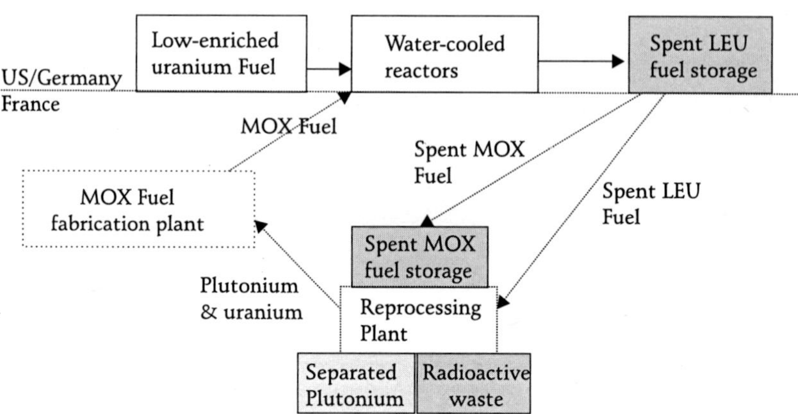


*Note*: Today, most countries store their spent fuel at their reactor sites and when the storage pools fill up, the older spent fuel is moved to dry cask storage. In France and the United Kingdom, most spent fuel is chemically 'reprocessed.' In France the separated plutonium and uranium are recycled once and then stored. The recycled plutonium is not separated out for further recycle because it contains a significantly increased fraction of plutonium isotopes that are not easily fissioned by the slow neutrons that mediate the fission chain reaction in water-cooled reactors.

Japan has built a $27 billion (¥2.2 trillion) plant designed to reprocess 800 tonnes of light-water-reactor-fuel annually (Figure 7.5). It was supposed to start up in 1997 but, has only reprocessed 400 tonnes so far because the melters, in which high-level reprocessing waste are supposed to be mixed into glass, are not functional. Recently, Japan's Atomic Energy Commission (JAEC) estimated that proceeding with the programme will increase the cost of electricity in Japan by $130-160 billion over the 40 year design life of the new reprocessing plant, relative to interim storage and then disposal of spent fuel in a deep underground repository.[6]

---

6. The JAEC estimated that reprocessing and plutonium recycle would increase the cost of nuclear electricity by ¥0.8-1.0 per kWh relative to simply disposing of the spent fuel, "Estimation of Nuclear Fuel Cycle Cost and Accident Risk Cost," 10 November 2011. The plan is for the plant to process 32,000 tons of spent fuel over its design life. Assuming an average of 50 megawatt-days per kg uranium fission energy release in the spent fuel, and a conversion of one third of the heat to electrical power, the extra cost incurred over the life of the reprocessing plant would be ¥10-13 trillion or $130-160 billion at current exchange rates.

**Figure 7.5**

*Rokkasho Reprocessing Plant, Japan*

*Note:* Japan's Rokkasho Reprocessing Plant will cost an estimated $2 billion (¥0.18 trillion) per year to operate and an estimated $23 billion (¥1.9 trillion) to decommission. It was designed to reprocess 800 tonnes of light-water reactor spent fuel per year, starting in 1997, but it has so far only reprocessed 400 tonnes and has been shut down since 2008.[7]

Because of the high cost of reprocessing and the fact that high-level reprocessing waste is not easier politically to dispose of than spent fuel, most countries are currently storing their spent fuel, pending the availability of a deep geological repository or a later decision to reprocess (Table 7.2)

Of the 31 countries with operating nuclear power plants:

- Eleven have never reprocessed their spent power reactor fuel. The United States, has not reprocessed since 1972.[8]

- Eleven countries sent spent fuel to France, Russia or the UK for reprocessing but decided not to renew their contracts. Ten of them decided to store their spent fuel instead. The 11th, Japan, decided to build a domestic reprocessing plant.

---

7. Slide 12 from a Japan Atomic Energy Commission PowerPoint dated 25 October 2011, *http://www.jnfl.co.jp/english/business/reprocessing.html*

8. The U.S. had a reprocessing plant from 1966-72 that operated under private ownership before being shutdown and transformed into a government-owned multi-billion dollar cleanup project that continues today and is expected to continue for at least a decade, U.S. Department of Energy, "West Valley Demonstration Project," *http://www.wv.doe.gov/*.

**Table 7.2**

*Policies Relating to Spent Fuel Reprocessing for the 31 Countries with Operating Nuclear Power Plants*

| Countries that reprocess or plan to (% of capacity) (GWe, [109 Watts]) | | Customer countries that have quit or plan to quit (GWe) | | Countries that have not reprocessed (GWe) | |
|---|---|---|---|---|---|
| China (20%) | 11.7 | Armenia (in Russia) | 0.4 | Argentina | 0.9 |
| France (85%) | 63.1 | Belgium  (France) | 5.9 | Brazil | 1.9 |
| India (≈50%) | 4.4 | Bulgaria (Russia) | 1.9 | Canada | 12.6 |
| Japan (100% planned) | 44.2 | Czech Republic (Russia) | 3.7 | Iran | 0.9 |
| Netherlands (in France) | 0.5 | Finland (Russia) | 2.7 | Mexico | 1.3 |
| Russia (10%) | 23.6 | Germany (France/UK) | 13.1 | Pakistan | 0.7 |
| Ukraine (5% in Russia) | 13.1 | Hungary (Russia) | 1.9 | Romania | 1.3 |
| UK (90%, not LWR) | 9.9 | Slovak Republic (Russia) | 1.8 | Slovenia | 0.7 |
| | | Spain (France, UK) | 7.6 | South Africa | 1.8 |
| | | Sweden  (France/UK) | 9.3 | South Korea | 18.7 |
| | | Switzerland (France/UK) | 3.3 | Taiwan, China | 5.0 |
| | | | | U.S. (since 1972) | 101.1 |
| Total (65%) | 170.5 | Total | 52.6 | Total | 146.6 |

Note:  Also shown is each country's nuclear generating and, for the countries that do reprocess, an estimate of the percentage of their capacity whose spent fuel is reprocessed

- Six reprocess. France and the UK have large operating reprocessing plants, although the UK plans to end reprocessing when its current contracts have been fulfilled. Japan has a large reprocessing plant that was supposed to be completed in 1997 but still is not operating due to technical problems; India operates three small reprocessing plants that process heavy-water reactor fuel; China and Russia operate pilot reprocessing plants.[9]

- The Netherlands and Ukraine have renewed their reprocessing contracts for small amounts of spent fuel that they are shipping to France and Russia, respectively.

---

9. Russia started construction on a large reprocessing plant near Krasnoyarsk in 1984 but abandoned the effort in 1989 because of lack of funding and public opposition. Russia now plans to build a new experimental reprocessing plant at that site, which has become Russia's central spent-fuel storage site.

## India Now has a Choice

For 30 years, from 1978 to 2008, India did not have access to the global uranium market. It was therefore, limited to domestic uranium resources. The exploitation of these resources is the monopoly of India's Department of Atomic Energy (DAE), which has reported extensive areas with favorable geology but only about 100,000 tonnes of 'reasonably assured resources', thus far.

In view of its ambitious plans for nuclear capacity expansion, its lack of access to the global uranium market, and the perceived limited domestic uranium resource in India, the DAE has emphasised the urgency of commercialising breeder-reactors.

In 2008, however, the Nuclear Suppliers Group lifted its ban on sales of uranium to India, and India immediately began to import uranium. Whereas, previously, the DAE expected that limitations on domestic uranium resources would limit its deployment of non-breeder reactors to 10 GWe, it now plans to add 40 GWe of light water reactors (LWRs) fueled with imported low-enriched uranium. However, it also plans to reprocess the spent fuel of these light water reactors to provide startup plutonium for breeder-reactors, which it expects to be the dominant reactor type in India, after 2030.

This is, of course, the same plan that France, Japan, Russia and the UK pursued with the result that today they have huge quantities of separated plutonium (Figure 7.6) but no plutonium-fueled breeder reactors.

## The Danger from High-level Liquid Waste

Reprocessing plants dissolve spent fuel to separate out the plutonium and uranium. The plan is then to 'vitrify' the residual liquid, containing fission products and minor transuranic elements by mixing it with molten glass to form a solid waste form that can be deposited into a deep underground geological repository. In practice, however, the vitrification process has proven difficult, with the result that large quantities of high-level liquid waste (HLLW) have accumulated on site.

Public information about HLLW is most available in the case of the UK. According to a report issued in 2000, by the UK Nuclear Installations Inspectorate (NII), the HLLW from the two operating reprocessing plants at Sellafield is stored in 21 tanks, 8 with volumes of 70 and 13 with volumes of 150 cubic meters each. About five of these tanks are currently in use. The amount of HLLW they contain is the equivalent of three to four years output of the reprocessing plants. The UK Office of Nuclear Regulation proposes that the quantity of dissolved fission products from oxide fuel stored in this form be limited to the equivalent of that in 2,000 tonnes of spent fuel, and that the total, including fission products from uranium metal 'Magnox' fuel be limited to the equivalent of 5,500 tonnes of spent fuel. This is roughly equivalent to a limit of 140 MCi of Cs-137, about 70 times the amount released by the Chernobyl accident.[10] The volume of HLLW at France's reprocessing plant at La Hague is comparable to that at Sellafield.[11]

In Japan, in a test run of the Rokkasho Reprocessing Plant, one of the two melters that mixes HLLW into glass, clogged in 2008. As a result, reprocessing operations at the $27 billion plant were shut down for four years while the engineers tried to understand the problem. In 2012, they tried the second melter, and it too clogged. As a result, 240 m3 of high-level waste from the reprocessing of about 400 tonnes of spent fuel remains in liquid form. Japan's Tokai Pilot Reprocessing Plant held a total of 380 m3 of HLLW as of the end of March 2011.[12]

India also appears to have been experiencing difficulties in setting up and operating a vitrification facility at the Kalpakkam

---

10. Assuming 4,000 MWt-days of fission per ton of Magnox fuel, 20,000 MWt per ton of Advanced Gas Reactor fuel, 3.2 Ci of Cs-137 produced per MWt-day and that, on average, the material is 10 years post reactor discharge.

11. As of 31, December 2009, at eight different locations at La Hague, there were 12 different stores of liquids classified by AREVA as HLLW: six containing fission products with a total volume of 667.9 m3 and six described as "effluent concentrates with a total volume of 509.2 m3", communication from AREVA to Mycle Schneider dated 6 April 2010.

12. The Japan Atomic Energy Agency's quarterly report, dated April 28, 2011 to the governor of Ibaraki prefecture on the operation of its nuclear facilities in the prefecture, for the period from January 1 to March 31, 2011.

reprocessing plant. As a result, the HLLW tanks there may contain up to 13 MCi of Cs-137.[13]

The difficulties of HLLW vitrification have provided an incentive for the designers of reprocessing plants to build high-volume HLLW tanks so that a delay in vitrification will not interrupt reprocessing. A proposed reprocessing plant at Gorleben in Germany was to have had a HLLW storage capacity of 7,000 cubic meters. A non-governmental critique, however, convinced the German regulatory authorities to rule in 1979 that, because of safety concerns, HLLW should be stored, if at all, only in small quantities in inherently-safe tanks. That ruling is consistent with best practice in chemical-plant design, which is to eliminate storage of hazardous intermediates. Unfortunately, that practice has not been followed for the reprocessing plants that have been built.

Therefore, the current practice in the reprocessing industry is to store large amounts of radioactive material in an acidic liquid form that requires constant cooling. This creates both economic and safety risks. The economic risks are that, if vitrification does not work at all, then reprocessing will make the spent-fuel disposal problem worse. And, if vitrification is delayed for years, then a choice is forced between accumulating hazardous liquid wastes, indefinitely, or idling a costly facility, as has occurred in Japan. If a radioactive release occurs inside a reprocessing plant, it also can idle the facility for years. That happened in 2005, at the Thermal Oxide Reprocessing Plant (THORP) plant where a pipe broke, a building was flooded with liquid high-level waste, resulting in plant closure for two years.

The greatest concern, however, is the possibility of a large air-borne radioactive release contaminating a significant part of a country, due to an accident or malevolent act.

---

13. There have been reports over several years about a vitrification plant being set up in Kalpakkam. In May 2008, Dr Srikumar Banerjee, currently the Chairman of the Indian Atomic Energy Commission, announced that the plant "will soon become operational", R. Prasad, "Plant to vitrify nuclear waste," *The Hindu*, May 15, 2008. However, thus far, it doesn't appear to have started operating. Therefore, the HLLW at Kalpakkam must have accumulated for the entire lifetime of the plant. The Kalpakkam plant had reprocessed an estimated 630-780 tonnes of spent fuel as of the beginning of 2010, International Panel on Fissile Materials, Global Fissile Material Report 2010, p. 121. Assuming an average cooling period of ten years, there would be about 17 kCi/tHM of Cesium-137.

As far as is publicly known, there has been no comprehensive assessment of these risks for any actual or proposed reprocessing plant. In 2000, the UK Nuclear Installations Inspectorate (NII) conducted a limited assessment for various possible release initiators including loss of tank-cooling and accidental aircraft crash. It was found that loss of tank-cooling could result in the contents heating up to boiling in a time as short as 12 hours. NII argued that, due to redundant arrangements for cooling, this was very improbable. For an aircraft crash into the HLLW storage building, it was estimated by the plant operator that the probability of perforating the building would be 1.7 per cent or less. Since, the likelihood of an accidental crash in the first place was estimated as less than one chance in a million per year, this scenario was not analysed further. (This was before the attacks of 11 September 2001, however, and terrorist attacks were not considered.)

## The Dangers of Plutonium Theft

About 250 tonnes of separated civilian plutonium have accumulated as a result of civilian reprocessing—about as much as was separated by the Soviet Union and United States for weapons during the Cold War (Figure 7.6). Most of the civilian plutonium is not weapon grade, but all of it is weapon useable. Eight kg is more than sufficient for the manufacture of a Nagasaki-type nuclear weapon.[14] The 250 tonnes of separated civilian plutonium is, therefore, sufficient for more than 30,000 Nagasaki-type nuclear weapons and should be guarded as a weapons material. In many cases, however, because of cost concerns and civilian cultures, civilian separated plutonium is not guarded to anywhere near the same security standards as military plutonium and is therefore, at greater risk of theft.[15]

---

14. IAEA, Safeguards Glossary (2001) p 23. For plutonium isotopic mixtures with high percentages of Pu-240, the yield in a Nagasaki-type design is likely to be less than the 20-kiloton yield of the Nagasaki bomb but, more than the detonation of 1000 tons of chemical explosive, J C Mark Science & Global Security, 4(1993) p.111

15. For a discussion of the exposure of civilian plutonium during transport in Europe, see, for example, International Panel on Fissile Materials, Global Fissile Material Report (2007) p. 17-18

**Figure 7.6**

*Separation of Civilions: Country-wise*

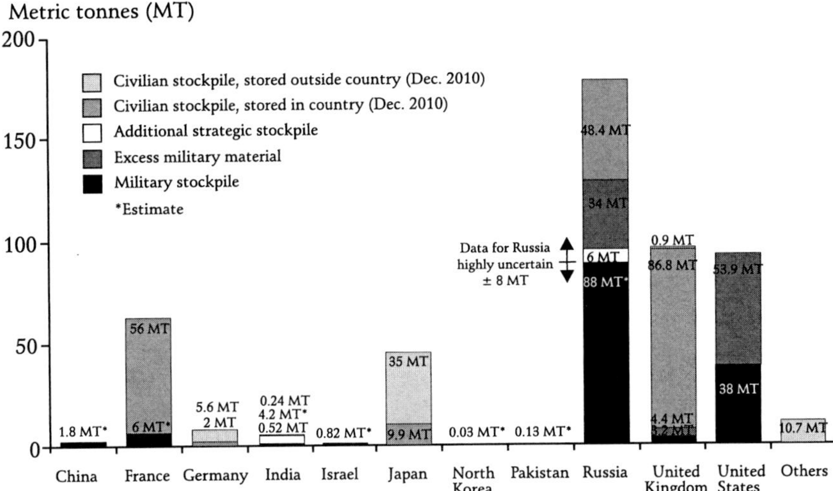


*Note:* Countries have separated as much civilian as weapons plutonium as a result of planning for the advent of breeder-reactors that have not materialised.

Civilian plutonium is mostly stored as $PuO_2$, which is exceedingly dangerous, if dispersed in the air. The inhalation of one gram by a large population would cause on the order of 10,000 cancer deaths.

## The Alternative: Spent Fuel Storage

Spent fuel storage is much less costly than reprocessing. That is why so many countries have abandoned reprocessing (Table 7.2). It is also a much more secure way of storing plutonium because of the self-protection provided by the gamma-emitting fission products—especially 30-year half-life cesium-137 (Figure 7.7). Finally, in the absence of breeder-reactors, it simplifies radioactive waste disposal. All the waste is confined in the spent fuel. In contrast, reprocessing creates multiple waste types, including ultimately the reprocessing plant itself.

**Figure 7.7**

*A Pressurised-Water Reactor Fuel Assembly and the Radiation Dose from it*

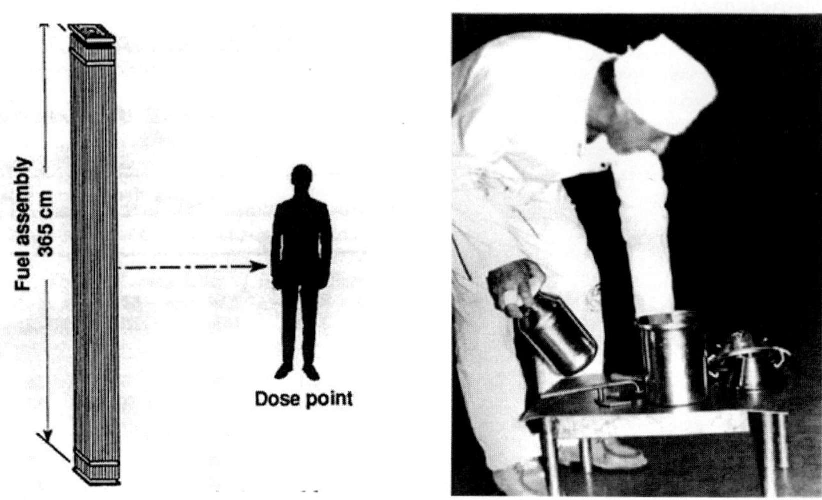


*Note:* The assembly, shown on the left, weighs about 500 kg and contains about 5 kg plutonium. Fifty years after discharge, a person standing at a distance of 1 meter, as shown, would absorb a potentially lethal gamma dose of 4 Sieverts within half an hour. The picture on the right shows a worker at Russia's Mayak pilot reprocessing plant placing a container holding 2.5 kg of plutonium in oxide form into an outer container (photographed during a visit by the author in 1994). There is very little penetrating radiation from separated plutonium. The main danger is from short-range alpha particles, if particles of $PuO_2$ are inhaled. For that reason, the worker is placing the container inside another to minimise the danger to exposure of leaking $PuO_2$.

Initially, when it is discharged from a reactor, spent fuel must be water cooled in a pool, next to the reactor. After several years of cooling, however, it can be shifted to a dry air-cooled cask. Figure 7.8 shows different forms of dry spent-fuel storage in several countries. At the upper right, are dry casks stored at the Fukushima Daiichi nuclear power plant that survived with their contents undamaged, the earthquake and tsunami and the consequent damage to the building housing them. There was so little concern about possible damage to the spent fuel inside, that Japan's public still is generally unaware of the presence of dry-cask storage of spent fuel at the plant.

**Figure 7.8**

*Dry Storage of Sent Fuel at Different Sites*

*Note:* *Top Left*: MacStor 400 Monoliths for the dry storage of CANDU heavy water reactor spent fuel, are in use in Canada and at South Korea's Wolsung Nuclear Power Plant. *Top right*, metal spent fuel storage casks at Fukushima Daiichi nuclear power plant that survived the accident with their contents undamaged, despite of the damage to the building sheltering them. *Lower left*, spent fuel stored in canisters in reinforced concrete shielding casks at the U.S. Connecticut Yankee Nuclear Power Plant, before it was decommissioned.[16] *Lower right*, shielding building containing dry cask stored spent fuel in at the Lingen Nuclear Power Plant, Germany.

## An Alternative Path for India

As described earlier, several countries, notably France, Japan and the United Kingdom, launched costly large-scale reprocessing programmes in anticipation of large-scale commercialisation of breeder-reactors, which eventually did not happen. In addition to wasting tens of billions of dollars of R&D funds, these programmes have unnecessarily exposed huge quantities of weapon-usable and radiotoxic plutonium oxide to possible theft and have created huge stocks of liquid high-level waste that are also potential terrorist targets.

---

16. *http://www.connyankee.com/html/fuel_storage1.html*

In light of this experience, India may wish to reconsider its own long term nuclear energy development strategy and shift away from an exclusive focus on breeder-reactors and plutonium separation. Now that it has access to the global uranium market, India also can reassess the economic justification of its plutonium breeder programme. What is the likelihood that liquid-sodium cooled breeder-reactors, with their complex closed fuel cycle, will prove to be reliable and competitive with water-cooled reactors on a once-through uranium fuel cycle?

If India decides to postpone its breeder commercialisation programme, it will always have the option to restart it in the future in the unlikely event that the economics change. The plutonium in the stored spent fuel will still be there.

Finally, if India decides to continue with its breeder R&D programme—for a time at least—it should minimise the costs and risks by:

- securing the separated plutonium, like the nuclear-weapon material it is;

- not separating plutonium before it is required; and

- Vitrifying high-level liquid waste immediately, so that large stocks do not accumulate.

## References

A Kakodkar Evolving Indian Nuclear Energy Programme – Rationale and Perspective (INAS,2008).

Beyea, J., Y. Lenoir, G. Rochlin, and G. Thompson (1979). *Report of the Gorleben International Review.*

Energy, Electricity and Nuclear Power Estimates for the Period up to 2050 (IAEA, 2011), Table 3.

Fetter, S. and F. von Hippel (1990). *Science & Global Security* 2: 2.

Fitts, R.B. and H. Fujii (1975). IAEA Bulletin 18(1): 19.

Hewlett, R.G. and F. Duncan (1974). Nuclear Navy: 1946–1962 University of Chicago Press pp 272–273.

*http://www.sangiin.go.jp/japanese/joho1/kousei/syuisyo/177/meisai/m177136.htm*

International Panel on Fissile Materials, Global Fissile Material Report (2011).

International Panel on Fissile Materials, Managing Spent Fuel from Nuclear Power Reactors: Experience and Lessons from Around the World (2011).

International Panel on Fissile Materials, Managing Spent Fuel from Nuclear Power Reactors (2011) Figure 2.1.

Lloyd, W.R., M.K. Sheaffer and W.G. Sutcliffe (1994). Dose rate Estimates from Irradiated Light-Water-Reactor Fuel Assemblies in Air, Lawrence Livermore National Laboratory, UCRL-ID-115199.

O M Saraev IAEA Fast Breeder Reactor database (1997).

Russia has one operating breeder reactor but it is fueled 17-26 per cent enriched uranium, Saraev, op. cit.

Saegusa, T., K. Shirai, H. Takeda, M. Wataru and J. Tani (2012). "Issues and Countermeasures for Long-Term Storage of Spent Fuel by Dry Cask", presentation at the US Nuclear Regulatory Commission Regulatory Information Conference (2012) session W9.

Safety and Security of Commercial Spent Nuclear Fuel Storage National Academy Press (2006) Figure C.2.

Szilard, Leo (1972). *The Collected Works of Leo Szilard: Scientific Papers* MIT Press pp 369.

The Storage of Liquid High Level Waste at BNFL Sellafield: An Updated Review of Safety (HM Nuclear Installations Inspectorate, 2000),pp 1-25, 1-26.

The Storage of Liquid High Level Waste at BNFL Sellafield, op. cit., pp. 2.26 and 2.140.

Uranium 2009: Resources, Production and Demand (OECD Nuclear Energy Agency and IAEA, 2009).

Uranium Intelligence Weekly (2010) pp 7.

UK Decommissioning Authority, Oxide Fuels: Credible Options (2011).

Uranium 2009: Resources, Production and Demand (OECD Nuclear Energy Agency and IAEA), pp 221.

UK Health and Safety Executive, Office of for Nuclear Regulation High Active Liquor Stocks Specification No. 793 (Project Assessment Report, 2011) pp 27.

8

BALDEV RAJ, P. CHELLAPANDI & S.C. CHETAL

# Science and Technology of Sodium Cooled Fast Spectrum Reactors and Closed Fuel Cycles in Three Stage Programme of India
## Accomplishments and a Way Forward

## Introduction

The current population of India in the year 2012, is about 1.16 billions, which would stabilise at the level of 1.5 billions by the year 2050, constituting one-sixth of the global population. India is on the rapid economic growth path. Per capita electricity in India is low (about 700 kWh/a), compared to 13000 kWh/a in USA and 8200 kWh/a in Organisation for Economic Cooperation and Development (OECD) countries. Hence, the demand for a rapid rise in the electricity generation capacity in the coming decades is beyond a matter of debate, for India to realise its dreams of a sustained growth in economy backed by strong industrial development. India is aiming to reach at least a per capita energy consumption of about 2400 kWh/a with eight per cent growth rate by 2031-32. This calls for the electricity generation capacity of about 778 GWe by 2031-2032 with the projected population of about 1.47 billion, by that time. It is worth mentioning that the present energy scenario is not satisfactory and the persistent shortage, unreliability and high prices for industries need to be eliminated, on priority, for India to be globally competitive and provide sustainable good parameters of life in education, healthcare, food, housing, infrastructures etc., to the citizens. Raising the electricity availability by about four times in the next 20 years, calls for a careful examination of all issues related to sustainability including relative abundance of available energy resources, diversity of sources of energy supply and status of technologies, security of supplies, self-sufficiency, security of energy infrastructure, effect on local, regional and global environment and demand management.

Modeling of scenarios coupled with policy initiatives along with launching of energy projects, aim at improving energy availability and improving transport and communication infrastructure. Further, the environmental implications of various resources are critically analysed by inter-governmental panel on climate change, chaired by the Prime Minister of India. Comprehensive assessment has established that improved and clean coal technologies, renewable (particularly solar and wind) nuclear energy based technologies are the scalable and suitable options for the country, from the comprehensive environmental considerations.

Taking cognizance of India's nuclear resource profile, Dr Homi Bhabha formulated a three stage nuclear power programme for achieving energy independency. The three stage nuclear programme, comprises of water reactors starting with pressurised heavy water reactors (PHWRs) in the first stage, sodium-cooled fast reactors (SFRs) in the second stage, and thorium-based reactors in the third stage. Nuclear energy, in India, is pursed with closed fuel cycle for maximising natural resources, and mitigating challenges of proliferation and high-level waste management.

The paper explains advantages of fast reactors and importance of developing closed nuclear fuel cycle through SFRs for large scale energy production, (though this is not the main scope of this paper). The scope of the paper is to highlight the importance of SFRs in the Indian Context, SFR programme in the country since, the inception starting with 40 MWt/13 MWe fast breeder test reactor (FBTR), to the current status and future programmes, is described (in brief). A detailed account of safety realisation through robust and comprehensive R&D is the main theme of this paper. Challenges and achievements in science and technology of SFRs related to safety are described with particular reference to 500 MWe capacity prototype fast breeder-reactor (PFBR), and associated fuel cycles.

## Advantages of Fast Spectrum Reactors (FSR)

In a nuclear reactor, fissile materials, such as $U^{233}$ or $U^{235}$ or $Pu^{239}$ are harnessed for producing energy through fission and converting kinetic energy of fission products to thermal energy. $U^{233}$ and $Pu^{239}$ are produced by conversion of the fertile materials, $Th^{232}$ and $U^{238}$ respectively. The ratio of fissile material produced and fissile material destroyed is termed as 'conversion ratio'. The conversion ratio, if greater than 1, is called breeding ratio. If $\eta$ is the number of neutrons produced per neutron absorbed, the necessary condition for breeding is: $\eta = 2 + x$, to account for one neutron for a new fission; one for a new conversion and x for leakages or parasitic captures.

Among various spectrum reactor types, for FSR in particular, $Pu^{239}$ is best for which the value of $\eta$, averaged over the energy spectrum is about 2.3. Hence, in a FSR, there is a net production of fissile material that can be used to fuel another reactor. This is the fundamental characteristic of fast neutron reactors. The conversion ratio in a typical 1000 MWe fast reactor core is ~0.72 and the same is 0.6 for pressured water reactor (PWR) of similar capacity. However, significant benefit is derived from the conversion in the blankets of the FSR. The conversion ratio rises to 1.12 with the additional conversion in the blanket. This is not the case with PWR, where surrounding the core with a blanket does not significantly change its conversion ratio of 0.6. This is primarily due to the low number of neutrons leaking out of the core. Hence, it is possible to modulate breeding ratio in the FSRs by enhancing the number of neutrons that are captured by changing blanket materials, thickness of the blankets, reflector thickness and materials, appropriately.

It is estimated that the ultimate accumulated long-lived radioactive isotopes, which need to be disposed in repositories, shall contribute less than 0.1 per cent of all the fission products, after multiple recycling in FSRs. There is no science or technological limitation of fuel recycling in FSR. The waste management burdens are reduced by about 200 times in terms of storage space and less than 700 years time span for the activity to become close to the background value. In view of possibility of higher operating temperatures in SFR, it is possible to have high thermodynamic efficiency. With the use of advanced materials for the fuel clad and wrapper, it is also possible to achieve, higher burn ups of the order of 200,000 MWd/tonne. These aspects lead to significant economic advantages. High burn ups in FSRs, in comparison to approx. 75,000 MWd/tonne for PWRs and 7000 MWd/tonne for PHWRs, has additional advantage of less thermal pollution to the environment, due to mining and front-end fuel cycle operations. It is obvious that FSR contributes to the reduction of the greenhouse effect ($CO_2$ emissions) compared to electricity generation using fossil fuels and even water-reactor systems, because of absence of significant mining in case of FSRs (as additional uranium requirement for running these reactors is small in comparison to water-reactors). These advantages were envisaged as early as in 1946 by Enrico Fermi, who explained in an explicit manner, the breeding principle and stated that the country, which will develop Liquid Metal based FSR technology, will lead the world, in the future.

## Importance of FSR in Indian Context

In view of availability of limited indigenous uranium in the country (~110 kt), the SFRs are important due to their capability of efficient utilisation of uranium. However, India has abundant thorium. It is often not realised that the fast reactors also play important and essential role to exploit the thorium resources for large energy production. Since, thorium cannot straightway be used to produce energy, as it has no fissile isotopes, it has to be bred to give fissile isotope ($U^{233}$) in blankets of Pu-U fuelled fast reactors or in water-reactors. This breeding is best done in fast spectrum reactors due to favourable availability of surplus neutrons for the purpose. However, early large scale introduction of thorium, hampers the growth in energy realisation by increasing the doubling time of nuclear fuel and hence, quantum of thorium for breeding in the second stage needs to be done in a planned strategic manner. Introduction of large scale thorium, without going through fast reactors, is not a desired strategy, since the installed power capacity with thorium and plutonium being used together in thermal reactors will lower the realisation of required amount of energy through indigenous nuclear resources.

Apart from the above advantages with SFRs, liquid metal technology and high temperature design inputs for fusion and high temperature reactor systems are the other tangible gains. The current strategy has the possibility to deliver electricity at competitive costs over long periods and lessen (considerably) the concerns of high level waste management and proliferation. Hence, SFRs are considered to be inevitable option for providing sustainable and environmentally acceptable energy systems and is the mainstay of nuclear power programme in India, from 2025 to 2050.

## Fast Reactor Programme in India

Indian fast breeder-reactor (FBR) programme was started by constructing a loop type fast breeder test reactor (FBTR), which is in operation, since 1985. With the PHWR programme well on the growth path and having established comprehensive expertise in SFR Technology through successful operation of FBTR over 27 years and comprehensive RD$^3$, India is now on a robust pathway for development of SFR- based, second stage of the programme. PFBR, was launched in October 2003, to realise the pathway for energy security. With construction nearing completion, PFBR is undergoing stage wise commissioning tests. It is envisaged that six more such units, with extensive additional RD$^3$ innovations in PFBR, and based on learning experiences, will be constructed by the year 2023. Subsequently, 1000 MWe SFRs using metallic core (with high breeding potential) will be constructed to realise the nuclear power rapidly. However, complete realisation of SFR technology involves many challenges, which are comprehensively addressed in this paper. The targets and strategies of SFR development are illustrated in Figure 8.1.

**Figure 8.1**

*FBR Programme in India*

- Indigenous design and Construction
- Comprehensiveness in development to design, R&D and construction
- High emphasis on scientific breakthroughs
- Synthesis of operating experiences
- Synthesis of emerging concepts (Ex. GENIV)
- Focus on national and international collaborations
- Emphasis on high quality human resources
- Creation of environment for enabling innovations
- Marching towards world leadership by 2025

MFBR
- 1000 MWe
- Pool Type
- Metallic fuel
- Serial constr.
- Indigenous
- Beyond 2025.

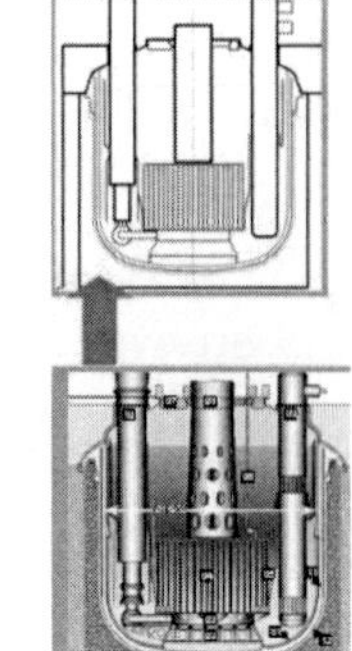

CFBR
- 500 MWe
- Pool Type
- UO2-PuO2
- 3 twin units
- Indigenous
- From 2023.

FBTR
- 40 MWt
- 13.5 MWt
- Loop Type
- PuC-UC
- Design: with French collaboration
- Since 1985

PFBR
- 1250 MWt
- 500 MWe
- Pool Type
- UO2-PuO2
- Indigenous
- From 2013.

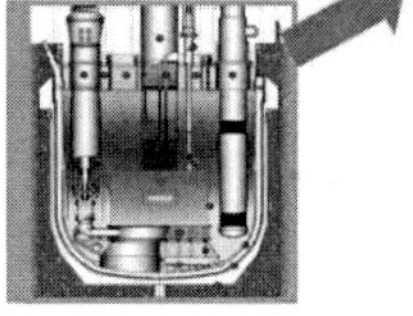

## FBTR and the Current Status

FBTR is a sodium-cooled, loop type fast reactor, fuelled with an unique high Pu mixed carbide fuel. It has two primary and two secondary sodium loops. Each secondary loop has two once through, serpentine type steam generators (SG). All the four SG modules are connected to a common steam-water circuit having a turbo-generator (TG) and a 100 per cent steam Dump Condenser (DC). The first criticality was achieved in October 1985, with a small core of 22 fuel subassemblies (SA) of MK-I composition (70% PuC + 30% UC), with a design power of 10.6 MWt and peak linear heat rating (LHR) of 250 W/cm. Progressively, the core was expanded by adding SAs at peripheral locations. Carbide fuel of MK-II composition (55%PuC+45%UC) was inducted in the peripheral locations in 1996. TG was synchronised to the grid for the first time in July 1997. LHR of MK-I fuel was increased to 400 W/cm in 2002. Eight high Pu MOX fuel SA (44% PuO2) were loaded in the core periphery in 2006. The reactor has so far been operated up to a power level of 20 MWt, and 18 irradiation campaigns have been completed. PFBR test fuel has been irradiated to a peak burn-up of 112 MWd/t. Indigenously developed unique Pu-rich mixed carbide fuel used in FBTR has performed extremely well, crossing a burn-up of 165,000 MWd/t. One of the important achievements is closing of the fuel cycle of FBTR. The FBTR fuel discharged at 155,000 MWd/t has been successfully reprocessed and refabricated. This is the first time that the Plutonium-rich carbide fuel has been reprocessed anywhere in the world. Towards designing and building future metallic fuelled test reactors, irradiation of metallic fuel pins is in progress. Further, the reactor life is to be extended by 20 years to serve as an irradiation facility for further development of technologies.

## PFBR and its Current Status

PFBR is a pool type reactor with 2 primary and 2 secondary loops with 4 steam generators per loop. Pool and loop type concepts were studied comprehensively, considering the associated merits and demerits, specific to medium-size reactors like PFBR. It was concluded that pool type reactor shall be our choice. The governing parameters meriting the choice are large thermal inertia that permits high thermal shock and higher structural reliability due to lesser number of critical welds associated and a compact layout of primary circuit components. It is also our perception that the complexities that are associated with the pool type of reactor such as pool hydraulics, manufacturing and handing of large dimensioned thin vessels with stringent tolerances, can be successfully met by the designers and our industry. Subsequently, it has been confirmed from detailed analysis, backed up with experimental validation and extensive 1:1 technology development exercise. The overall flow diagram, comprising primary circuit housed in reactor assembly, secondary sodium circuit and balance of plant (BoP), is shown in Figure 8.2. The nuclear heat generated in the core is removed by circulating sodium from cold pool at 670 K to the hot pool at

820 K. The sodium from hot pool, after transporting heat to four intermediate heat exchangers (IHX), mixes with the cold pool. The circulation of sodium from cold pool to hot pool is maintained by two primary sodium pumps and the flow of sodium through IHX is driven by a level difference (1.5 m of sodium) between the hot and cold pools. The heat from IHX is in turn transported to eight steam generators (SG) by sodium flowing in the secondary circuit. Steam produced in SG is supplied to turbo-generator. In the reactor assembly (Figure 8.3), the main vessel houses the entire primary sodium circuit including core. The sodium is filled in the main vessel with free surfaces, blanketed by argon. The inner vessel separates the hot and cold sodium pools. The reactor-core consists of about 1757 sub-assemblies, including 181 fuel sub-assemblies. The control plug, positioned just above the core, houses 12 absorber rod-drive mechanisms. The top shield supports the primary sodium pumps, IHX, control plug and fuel handling systems. PFBR uses mixed oxide with depleted uranium and approximately 30 per cent Pu oxide as fuel. For the core components, 20 per cent cold worked D9 material (15% Cr- 15% Ni with Ti and Mo) is used to have better irradiation resistance. Austenitic stainless steel type 316 LN is the main structural material for the out-of-core components and modified 9Cr-1Mo (grade 91) is chosen for SG. PFBR is designed for a plant life of 40 y with a load factor of 75 per cent.

Bharatiya Nabhikiya Vidyut Nigam Limited (BHAVINI), a government company was established for the construction and operation of FBRs in the country and PFBR is its first project. The construction of PFBR is almost completed, and is being commissioned in steps, to be operational in 2013. The nuclear steam supply system components have been manufactured successfully by the Indian Industries, based on the experience gained through technology development and including the feed back from the in-sodium testing. Manufacture of all reactor assembly components, such as safety vessel, main vessel, inner vessel, thermal baffles, grid plate, primary sodium pipes, roof slab, large and small rotatable plugs, IHX, pumps, fuel transfer machines have been completed exceeding the stringent tolerance requirements envisaged by designers. The safety vessel, incorporated with delicate thermal insulation panels, main vessel, core catcher, grid plate, inner vessel and roof slab have been erected successfully, meeting all the specified erection tolerances.

**Figure 8.2**

*Schematic of PFBR Flowsheet*

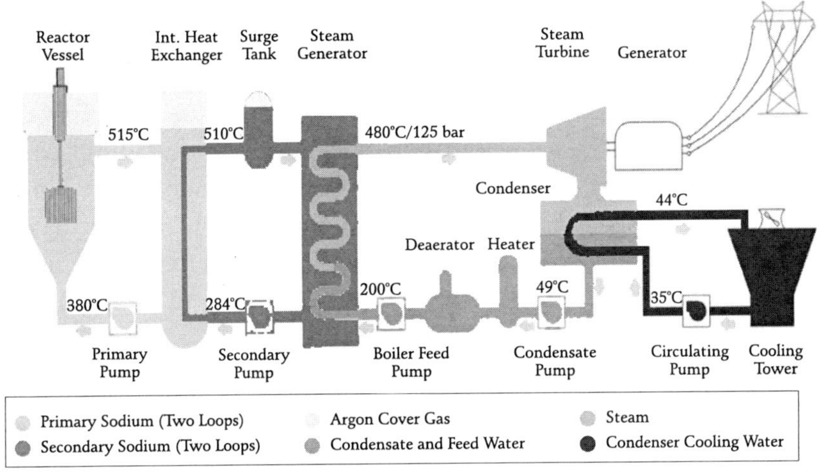


**Figure 8.3**

*PFBR Reactor Assembly Components*

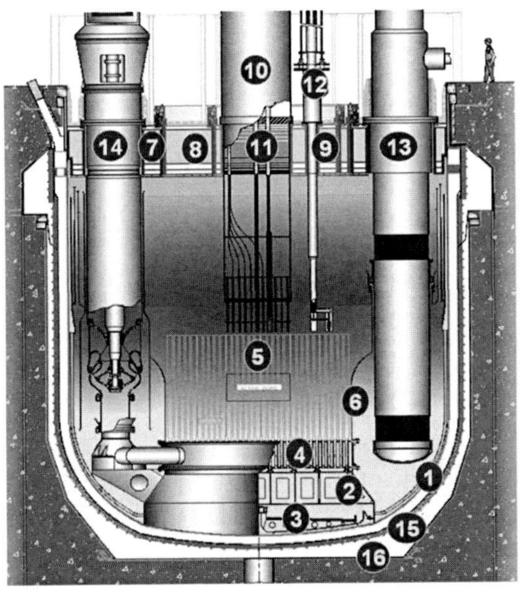


01. Main Vessel
02. Core Support Structure
03. Core Catcher
04. Grid Plate
05. Core
06. Inner Vessel
07. Roof Slab
08. Large Rotable Plug
09. Small Rotable Plug
10. Control Plug
11. Control and Safety Rod Mechanism
12. In-Vessel Transfer Machine
13. Intermediate Heat Exchanger
14. Primary Sodium Pump
15. Safety Vessel
16. Reactor Vault

## Design and Development of Future SFRs

As a follow-up on PFBR, it is planned to construct six 500 MWe commercial fast breeder reactors (CFBR) during 2015-2023 with innovations to enhance safety and economy through improved materials and manufacturing, and reduced construction time and commissioning period. One twin unit (2x500 MWe) will be constructed at Kalpakkam adjacent to PFBR. In CFBR, MOX (Mixed Oxide) fuel and two loop concept would be retained. Towards improving economy, twin unit concept, optimum shielding, use of 304 LN in place of 316 LN for cold pool components and piping, 3 SG modules per loop with increased tube length of 30 m (PFBR has 4 modules per loop with 23 m length), 85 per cent load factor, 60 years design life, reduced construction time (5 y) and enhanced burn up (upto 200 GWd/t to be achieved in stages) are being considered. Further, significant improvements would be introduced in the reactor assembly design. The improved design concepts under consideration are: (i) compact and symmetric welded grid plate without fuel transfer post; (ii) inner vessel having single curved redan integrated with fuel transfer post; (iii) thick plate rotatable plugs; (iv) control plug integrated with small rotatable plug; (v) torus shaped thick plate roof slab; (vi) torus support skirt for reactor assembly with optimum support location to minimise the seismic moments; (vii) safety vessel made of carbon steel integrated with reactor vault liner; and (viii) simplified fuel handling scheme with elimination of inclined fuel transfer machine. The improved design concepts have indicated significant economic advantages: material inventory reduction by $\sim$ 25 per cent, simplified fuel handling scheme and reduced manufacture time as well as enhanced safety. The innovations are being tested through systemic research, technology development, testing and evaluation, etc.

## Safety Features of SFR

The large margin between the operating sodium temperature and the boiling point of sodium, can accommodate significant temperature rise in the event of mismatch of heat generation and heat removal. High thermal conductivity, low viscosity and large difference between the temperature of hot sodium at 820

K and ambient air at 310 K, coupled with significant variation of sodium density with temperature, permits decay heat removal through natural convection mode. Pool type concept provides large thermal inertia and hence more time to the operator to act in case of exigencies during reactor operation. Temperature and power coefficient of reactivity are negative. So, in the event of disturbances in primary and secondary sodium flow or feed water flow, the reactor stabilises to a new power level even without the corrective action of the operator.

A few general concerns on safety of reactors have been addressed with defence in depth strategy. The core is not in its most reactive state, in the sense that the change of core geometry, specifically the core compaction, may lead to positive reactivity. Further, coolant voids in the central part of core would also cause positive reactivity. Fast reactors call for rapid control and shutdown systems. These are misconceptions. The solution to the reactor kinetic differential equation with the consideration of the delayed neutrons is represented in Figure 8.4. The periods, are shown on the ordinates; in these periods, the power of unregulated reactor increases by a factor of 3, if the appropriate reactivity is inserted without considering any feedback effects, the controlled reactors are practically not distinguishable from each other, from the reactor period point of view, as long as the reactivity is lower than about 1 \$. The periods are so long that a control is quite possible with technological robust mechanical regulating units. With the reactivity higher than 1 \$, the prompt neutrons determine the time-behaviour of the reactor (super-prompt criticality). The periods are too short in both reactor types, in order to be able to interfere effectively with a fast shutdown. Thus, every reactor must be designed so that a super-prompt-critical condition cannot occur and in all cases, this can be considered as a hypothetical event. If a super-prompt critical excursion is arbitrarily postulated, it is shown that it is limited highly by feedback effects. In this case, the doppler coefficient plays a prominent role. It all depends on the neutron absorption capability of $U^{238}$, available in the fuel material and increases as the fuel temperature increases. It has been shown that the excursion-energy is small through an extensive reactor experiment, called South-west Experimental Fast Oxide Reactor (SEFOR), in USA.

**Figure 8.4**

*Reactor Period versus Reactivity*

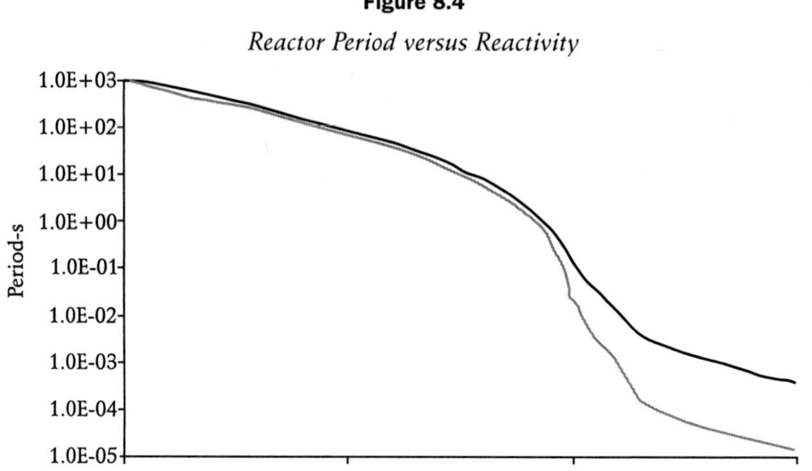


These apart, there are only a few initiators, which can introduce positive reactivities in SFR core *viz.,*: (i) gas entrainment; (ii) oil entry; and (iii) coolant voiding. Regarding gas entrainment effects, though isolated gas bubbles of relatively small sizes distributed in the core may be a possible scenario, they do not cause unacceptable reactivity variations. Further, it is also impossible to conceive a scenario, where a very large quantity of gas bubble (a few hundreds of litres) passes through the core that can cause significant reactivity change. Hence, gas entrainment is not an issue in fast reactor safety. However, as a defence-in-depth, the design features (mechanical seals for the intermediate heat exchangers, porous plates attached with the reactor internals in the vicinity of sodium free levels) are incorporated to avoid any possible gas entrainment. Quantity of oil used especially for the pump bearing, is minimised to the extent, that accidental oil leak is not of concern to the safety. Again a defence-in-depth, oil catch pots are provided below the oil tanks to prevent the oil fall into the reactor sodium. Apart from this, oil free bearings (ferro-magnet bearings) are conceived for the future design of pumps to completely eliminate the oil entry issue in the core of the reactor.

## Science, Engineering and Technology of SFRs w.r.t Safety

Science of SFRs involves understanding of unique fuel and structural materials under high temperature, sodium, irradiation environments over the long reactor life, sodium chemistry, aerosol behaviour, sodium fire and sodium water reactions, special sensors for sodium applications (detection of water leaks in steam generator, sodium leaks, purity measurements, level detectors), thermal hydraulics and structural mechanics (turbulences, instabilities, gas entrainments, thermal striping, stratifications, ratcheting, etc.). Various failure modes, which form the basis for the structural design are depicted in Figure 8.5. In the domain of engineering, the design for components at high temperature and long life, design and testing of mechanisms operating in sodium and argon cover gas space, design to accommodate sodium leaks, sodium-water reactions, seismic design of interconnected buildings, components and thin shells with fluid-structure interactions, in-service inspections (ISI), repair of reactor internals, development and qualification of high temperature fission chamber, etc., are a few typical examples of design challenges. Technology of SFRs demands indigenous development of codes, manufacture of large dimensioned welded thin shell structures made of austenitic stainless steel petals with close tolerances, machining of large dimensioned and tall slender components with stringent tolerances (grid plate, absorber rod drive and component handling systems), fabrication of large size box structures with controlled distortions, hard facing technology with special materials, development of inflatable seals, large size bearings, indigenous development of complex and sophisticated manufacturing technologies and reprocessing and waste management science and technologies.

**Figure 8.5**

*Typical Analyses to Comply Design Codes*

*Structural Mechanics*

- Creep-fatigue damage assessment
- High thermal cycle effects
- Flow induced vibration
- Buckling of thin shells
- Seismic Analysis
- Fracture mechanics
- Fast transient fluid structure interaction

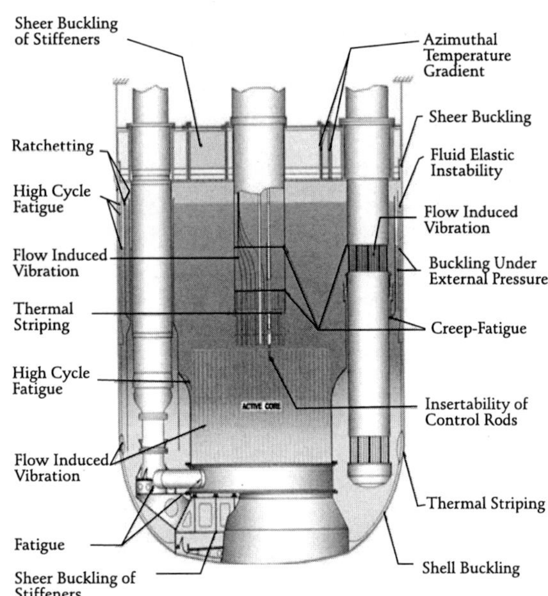


## Highlights of Safety Related Studies

### Thermal Hydraulics Studies towards Core Monitoring

The core is monitored by functionally diverse sensors, *viz.*, a variety of neutron detectors, flow meters, temperature monitors, etc. These parameters provide protection against transient over power, transient under cooling and anomalous reactivity addition events. Monitoring of sodium temperature is done by providing three (or) four fast response thermocouples mounted on the central canal plug to monitor central fuel sub-assembly sodium outlet temperature. Temperature of each fuel sub-assembly is measured through fast response thermo couples with full back up. Detailed thermal hydraulics analysis is carried out to determine flow and temperature distributions below the core cover plate to quantify the dilution in the measured sodium outlet temperatures, due to manufacturing and erection tolerances, and thermal and irradiation induced bowing of the sub-assemblies, to obtain an optimisation of number of flow zones (more number results in complex administrative control for subassembly (SAS) loading and lesser number calls for higher pump capacity), to derive accurate hot spot temperatures in fuel, clad and coolant.

## Assessment of Performance of Safety Grade Decay Heat Removal (SGDHR).

Decay heat is about 1.5 per cent and 0.7 per cent of nominal power, respectively, 1 h and 1 d after reactor shutdown. If off-site power is available, decay heat removal is through the normal heat transport system, i.e., through steam generators and steam-water system. The system is known as operation grade decay heat removal system. In case of loss of off-site power, loss of secondary circuit or steam water circuits, the decay heat is removed through 4 independent safety grade decay heat removal (SGDHR) loops. The SGDHR operation is robust in even most demanding envisaged conditions of accident. Each loop consists of a decay heat exchanger (DHX) of capacity 8 MWt with tube side linked to an intermediate sodium circuit, which is connected to sodium-air heat exchanger (AHX). The ultimate heat sink is air. The layout of SGDHR circuit ensures decay heat removal by natural convection in primary sodium, intermediate sodium and air side. Two dampers of diverse design are provided at inlet and outlet of AHX and two diverse designs of DHX and AHX are provided to enhance reliability. Diesel and battery power are also provided to drive the primary pumps at 15 per cent of the speed for conditions of off-site power failure and station blackout conditions as a defence in depth approach. Reliability analysis is carried out by fault tree method including common cause failure between redundant non-diverse components and systems. Hot sodium ensuing out of the core ($\sim$550$^\circ$C) reaches top surface of the pool and enters the DHX, gets cooled to $\sim$350$^\circ$C and leaves the DHX at the bottom of the hot pool. This cold sodium has a strong potential to remove significant amount of the decay heat, while it flows over the outer surfaces of the fuel sub-assemblies. The heat removal capability of this flow, known as inter wrapper flow, has been established based on a combined 'system level–multidimensional' thermal hydraulic model (Figure 8.6). Based on this study, it is found that the temperature limits of the core are satisfactorily met even under strong cyclones through natural convection.

**Figure 8.6**

*Analysis of SGDHR System*

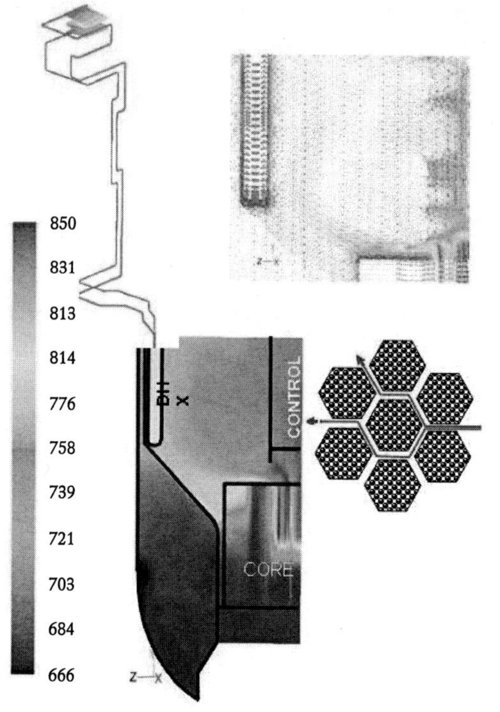


Analysis with multi-dimensional model for pools with inter-wrapper space (StarCD) and 1-D model for equipments and piping (DHDYN).

Availability any two circuits for 7h and one circuit subsequently with primary circuit under natural convection is sufficient to limit the temperatures below category 4 limits

## Reactor Safety under Seismic Loadings

The geometrical characteristics, particularly large size thin walled shell structures, large amounts of sodium, mass distributions and boundary conditions enhance the effects of seismic events. The main vessel carries about 2500 tonnes contributed mainly by sodium (1150 tonnes), core (550 tonnes) and self weight of main vessel and its internals. The weight of top shield along with the components supported by it is about 1200 tonnes. Thus, the reactor assembly weighs approximately 3700 tonnes, which is hanging from the top support. Further, there exists relatively thin annulus of liquid between: (i) inner vessel to inner baffle; (ii) inner baffle to outer baffle; and (iii) outer baffle to main vessel. The annulus gap to diameter ratio is W/D 100, which contributes in a significant manner to integrity of adjacent shells. The existence of large free fluid surfaces is the source of sloshing phenomena during normal operation as well as seismic events. Due to these features, the reactor assembly components have their fundamental natural frequencies less than 10 Hz, over which seismic floor responses show maximum dynamic amplifications. Hence, seismic events impose high dynamic forces, even though mechanical loadings (self weight and hydrostatic pressure heads) are low under normal operating conditions. For the large size thin walled shells (D/h >500), buckling is the dominant failure mode under dynamic forces developed during seismic events.

The reactor safety is ensured by ensuring the following design criteria:

- The reactivity insertion because of core compaction due to horizontal displacements and relative vertical displacements between the absorber rod and fuel subassembly, should be less than 0.5 $.

- It should be possible to insert control rods during safe shutdown earthquake (SSE).

- The drop time of absorber rods should be less than 1 s.

- Primary pumps should not seize.

- Sodium ejection to the reactor containment building (RCB) through the top shield penetrations is to be prevented.

- The stresses at the components should be less than the allowable values as per the RCC-MR, to ensure that there is no risk of buckling and loss of structural integrity.

In order to demonstrate that the reactor assembly components meet the above mentioned safety criteria, advanced numerical and experimental investigations are carried out. Typical results are highlighted in Figure 8.7.

**Figure 8.7**

*Investigations to Comply Seismic Design Criteria*

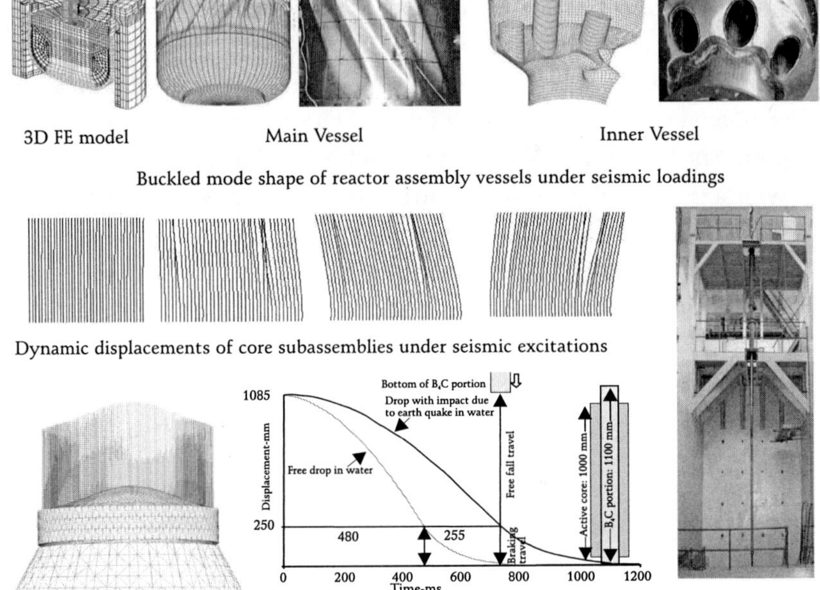



## Severe Accident Analysis

In order to improve understanding of the whole core melt down accident, it is necessary to understand the molten fuel coolant interaction (MFCI) effects. Towards this, sodium fuel interaction facility (SOFI) has been built with an objective to investigate molten uranium and uranium oxide interactions with sodium coolant experiments. The phenomenon chosen for the study are: solidification, fuel fragmentation, debris relocation and settlement behaviour on core catcher. The data collected are used for developing numerical simulations of scenarios during pre-disassembly and postaccident phases. Further, in order to simulate the heat transfer characteristics of debris settled on core catcher, specific tests are conducted to get data on debris–bed parameters.

## Sodium Safety

Though sodium is an excellent coolant in terms of its physical and neutronic properties, its chemical properties, especially last exothermic chemical reactions with the ambient air and water are of concern. Apart from many design measures to mitigate its effects, novel techniques are developed to address the related safety issues in a comprehensive manner. For achieving these objects, it is necessary to develop fundamental understanding of the chemical reaction scenarios, especially for the sodium spray fire scenarios (more damaging). Sodium spray fire experiments were carried out in MINA (MIni NA) facility to understand the combustion behavior of sodium droplets at 500°C and characterise the sodium aerosol produced during sodium fire. The variation of flame diameter of a single droplet was recorded using a high speed camera and it is observed that the combustion follows shrinking model concept. The burning rate of sodium droplets and aerosol mass distributions are theoretically calculated and compared with experimentally measured values. Samples of lime stone concrete and sodium resistant tiles were tested in MINA sodium spray fire facility. Temperatures at different locations, pressure, release of hydrogen and oxygen gas were monitored online during the experiments. The rate and extent of water release from concrete was determined by thermogravimetric analysis. Preliminary analysis of test results have revealed that when hot sodium at 500°C reacts on cold concrete block, it monotonically cools down with low degree of interaction. Depletion of water at the surface of the concrete was observed to be about 0.2 per cent in sodium resistant concrete. Tests are carried out to understand the scenario of cable fire following sodium spray fire. Imaging of sodium fire using high speed camera shows that cable fire follows the sodium fire and burns for longer period than sodium. Online and offline measurement of gases evolved during the combustion was also carried out. Analysis of the residue obtained as a result of sodium-cable interaction shows that chlorine gas released during combustion reacts with burning sodium to form sodium chloride. In addition to temperature rise, there is an increase of pressure (1.18 bar) inside the quartz chamber due to the gases released during cable fire. A few highlights of the results are presented in Figure 8.8.

**Figure 8.8**

*Safety Studies Related to Sodium*

NA spray fire scenarios and burning characteristics

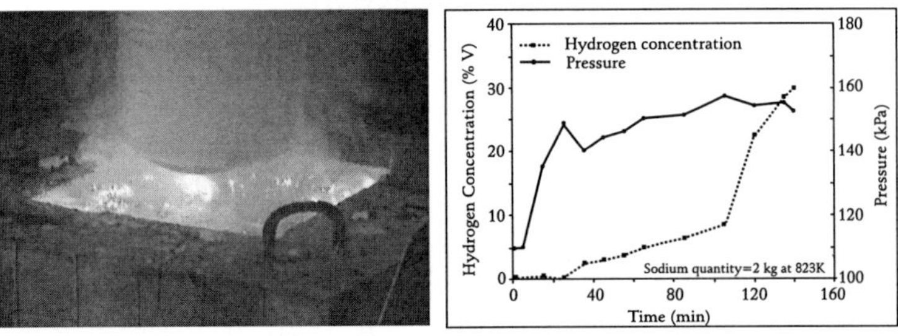


NA concrete interactions (Hydrogen production)

NA fire followed by cable fire (Top shield platform)

## Science and Technology of Sensors

Robust measurements are crucial for ensuring structural integrity and early detection and mitigation of anomalies during operation. Sensors have been developed for measuring ultra trace levels of dissolved hydrogen, carbon and oxygen contents in liquid sodium, in addition to monitoring the level, flow velocity etc. in sodium circuits. Detection of an event of a steam leak at the steam generator of the fast reactor is needed to avoid escalation and propagation of the leak, which could result in large maintenance issues. Based on detailed investigations on thermochemical and electrochemical investigations of the candidate, hydride ion conducting electrolyte systems, a sensitive electrochemical hydrogen sensor has been developed for detection of the steam leak. This sensor operating at 723 K uses $CaBr_2$-CaHBr as solid electrolyte and can measure down to 50 ppb of hydrogen in sodium with capability to detect a change in hydrogen level of 15 ppb at this concentration level. Tests in FBTR and also in large experimental sodium loops for studying long term performance, have shown the sensor to be stable and reliable with short response times. The sensor is unaffected by process parameters. The sensor and the measurement instrumentation are simple, cost effective and need low maintenance in contrast to the existing leak detection technology using high vacuum and mass spectrometry. For detecting the steam leaks during the start up or low power operating conditions of the reactor, a sensor system that measures continuously hydrogen levels in argon cover gas has been developed. This uses a long thin-walled nickel tube in the form of a coil, which is positioned in the cover gas plenum and measurements maintained at 773 K and through which high purity argon is made to flow. The hydrogen diffusing through the nickel coil into the argon stream is measured by a thermal conductivity detector and a sensor using thin film of tin oxide. This system has capability to measure from a few to several thousands of ppm of hydrogen in the cover gas. An electrochemical carbon sensor for measuring ppm levels of dissolved carbon in sodium has also been developed for continuous monitoring of carbon potential of sodium and to detect any hydrocarbon oil leak from the shaft cooling assembly of the centrifugal pumps used in sodium circuits. Online monitoring of dissolved oxygen in sodium is needed since, high oxygen levels result in enhanced corrosion and mass transfer in the structural steels used in the sodium circuits. A sensor using oxide ion conducting yttria doped thoria (YDT) ceramic has been developed for meeting this requirement.

Another important developments for the robust non-destructive examinations as well as in-service examinations is the automated eddy current imaging for location of weld, e.g., in the main vessel (stainless steel) of PFBR project and determination of its centreline for examination by ultrasonic testing. Identification of weld centre line is necessary for fixing the required skip distance and scan ranges for ultrasonic testing. In the case of austenitic stainless steel welds, the weld region with complex microstructures having delta ferrite and other phases, possess different electrical conductivity and permeability as compared to the base metal and this helps in eddy current testing to identify the weld centre line. During raster-scan imaging of welds, the weld region produces large amplitude changes in coil impedance and the grey level image, and 3-D profile image give distinct information of the weld. The accuracy of detection of the weld centre line has been found to be ± 0.1mm. As the inspection would be carried out at 150° C, a high-sensitive differential eddy current sensor with capability for operating at 200° C and for non-contact detection with a lift-off of 10 mm has been developed. This sensor would be integrated into sensor head of an inspection robot for automated and remote inspection of main vessel.

One of the limitations of sodium is its opaqueness, which poses a major problem for the in-service inspection of components. In this regard, indigenous development of vibration measurements systems and under-sodium viewing play a key role. The under-sodium scanners are essential for the successful fuel handling operations without having any risk of damage to fuels bundles as well as fuel handling parts. Fuel handling problems have been experienced in FBTR, JOYO (Japanese reactor) and BN type reactors (Russian). Robust systems developed at Indira Gandhi Centre for Atomic Research (IGCAR) are state of art and are validated through rigorous testing and evaluation.

## Manufacturing Technology

Structural integrity of components, apart from material choice, robust design, analysis and testing, depends strongly on the quality of manufacturing and inspection. Manufacturing deviations with respective to geometry and weld quality should be controlled in a stringent manner respecting international standards such as RCC-

MR (French construction codes). Thus, for the safe and reliable operation of SFR components, best manufacturing practices need to be followed.

Development of large size bearings, inflatable seals, high temperature fission chambers, manufacture of large size thin walled vessels made of stainless steel with tight form tolerances, machining and assembly of grid plate and steam generators with close tolerances, are some of the challenging issues that have been successfully resolved through detailed technology development exercise. In particular, for the large diameter thin vessels, the major manufacturing challenges are: the basic plates should not have any defects such as laminations ( best melting practices and high quality control is essential), large lengths of welds for integration of individual petals, stringent control on the manufacturing deviations, such as form tolerances ($<\frac{1}{2}$ thickness), verticality & horizontality ($<\pm2$ mm), high quality welds and low residual stress need to be achieved without any heat treatment. In order to build unique capability in the Indian industries and also assess manufacturing tolerances that can be achieved by the industries, elaborate manufacturing technology development works were undertaken prior to start construction of PFBR.

## Fast Reactor Fuel Cycle Technology

Nuclear fuel cycle for fast reactors has been a subject of intense R&D in a number of countries. Figure 8.9 depicts the generic closed fuel cycle concept. Besides, development of indigenous technologies for the various steps in the fuel cycle such as fuel fabrication, reprocessing and waste management, R&D in fuel cycle is important for establishing a robust fuel cycle for fast reactors that would not only address economy, safety and proliferation resistance, but also address the main concerns to the public, which relate to the presence of high concentration and quantities of plutonium and fission products in the reprocessing and waste management steps for fast reactors.

**Figure 8.9**

*Fast Reactor Fuel Cycle*

- Optimal and sustainable utilisation of valuable fissile material resources.

- Reduction in nuclear waste generation.

- Essential for growth of fast reactor programme.

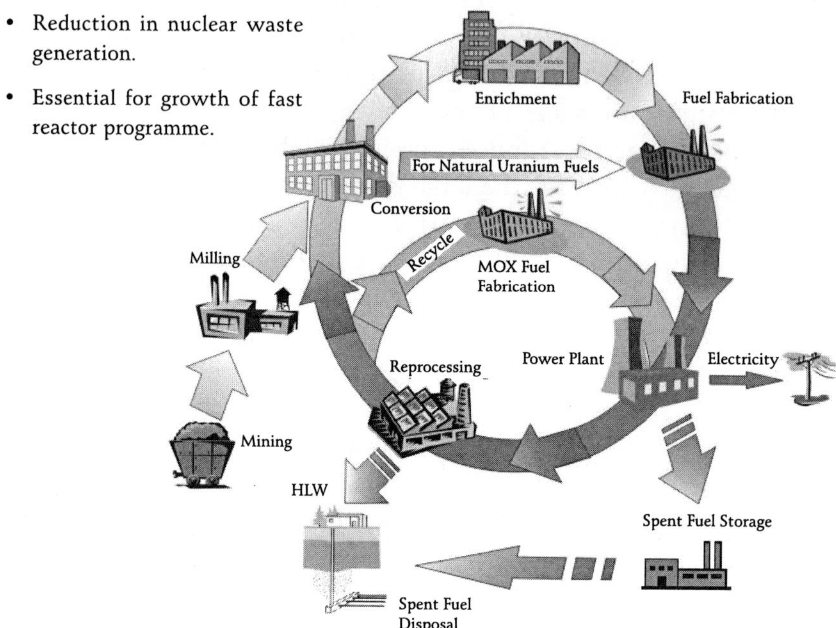


Reprocessing plants have been setup in a few countries for processing the oxide fuel from thermal reactors. Specific facilities and plants have been established for reprocessing the fast reactor fuel. These are located in Dounreay, UK and La Hague, Marcoule in France, Tokai Mura in Japan, Ozersk in Russia and Compact Reprocessing of Advanced fuels in Lead (CORAL) facility at Kalpakkam, India. The plants along with their capacities and the fast reactors they are catering to, are given in Table 8.1. Reprocessing of irradiated fuel, and especially the blanket assemblies, is fundamental to the SFR fuel cycle. Typically the recovered plutonium from aqueous reprocessing is incorporated into the core as MOX fuel and any surplus deployed elsewhere. However, with the transition from core and blanket designs to integrated core designs, it is likely that spent fuel will be reprocessed using electrometallurgical processes (pyro-processing) and plutonium will not be separated but will remain with some highly radioactive isotopes. Pyro-processing possesses several advantages for fast reactors, which greatly simplify waste management. The technology of Pyro-processing needs maturity and for the purpose, comprehensive R&D is in place. Aqueous reprocessing has been developed to a mature stage and the challenge relates to scale up of the technology in a well planned and phased manner.

**Table 8.1**

*World Scenario of Fast Reactors and Fast Reactor Fuel Reprocessing Plants*

| Country | Fast Reactor | Rating (MWe) |
|---|---|---|
| France | Rapsodite | 40 MWt |
| | Phenix | 250 |
| | Super Phenix | 1200 |
| | ASTRID | 600 |
| U.K. | DFR | 15 |
| | PFR | 250 |
| Russia | BOR60 | (12) |
| | BN350 | 130 MWe+Desalination |
| | BN600 | 600 |
| | BN800 | 870 |
| | BN1200 | 1200 |
| U.S.A. | EBR II | 20 |
| | FFTF | 400 MWt |
| Japan | Joyo | 140 MWt |
| | Monju | 280 |
| China | CEFR | 23 |

| Location | Capacity by | Total qty processed, t | Remarks |
|---|---|---|---|
| Dounreay, UK | 10 | 14 | DFR, PFR |
| La Hague (AT1) | 1 kg/day | | Rapsodie |
| La Hague (UP2) | 400 | 10 | Phenix also |
| Marcoule (SAP) | 5 | 8 | Rapsodie & Phenix |
| Tokai Mura, Japan | | 18 | Joyo |
| Ozersk (Chelyabinsk)-RT1 | 400 | | FBR also |
| Kalpakkam (CORAL) | 12 kg/year | | FBTR |

*Source*: Cacuci, Dan Gabriel (ed.) (2010). "Chapter 21: Sodium Fast Reactor Design: Fuels, Neutronics, Thermal-Hydraulics, Structural Mechanics and Safety", in *Handbook of Nuclear Engineering*. Springer Science+Business Media LIC. pp.2321-710.

**Fuel Choices**

Current choice of the fuel is mixed oxides of uranium and plutonium. In India, the MOX fuel has been chosen for the prototype and first CFBRs. The choice is mainly based on the vast international experience and the established technologies for the front end and back end of fuel cycle. FBTR employs the mixed carbides with high level of Pu enrichment at 70 per cent and subsequently with an additional variant with 55 per cent enrichment. PFBR uses MOX fuel at two enrichment levels with 21 per cent and 28 per cent PuO2. For the future on a long term, India would adopt metallic fuel towards getting a higher growth rate with high breeding ratio. In this, both binary and ternary alloys of metallic U and Pu with Zr as the alloying element are considered for comprehensive R&D programme.

*Fuel Cycle Activities in India*

The major goals in the development of reprocessing technology in India are:

- Processing of high burn-up fuel (current target: 200,000 MWd/tonne).

- Processing of discharged fuel after optimum cooling period.

- Recovery of minor actinides and societally valuable fission products to reduce radio-toxicity of waste and reduce time period over which it should be monitored.

In carrying out the developmental activities, several challenges are to be overcome for which, concerted efforts are undertaken. To name a few challenges: (i) building robust experience with high burn-up and short cooled fuels; (ii) development of robust centrifugal contactors; (iii) processing of fuel with high Pu concentration with emphasis on high recovery; (iv) high reliability in process as well as equipment to enable operation at name plate capacities and with high efficiency; (v) identification of new waste matrices that can incorporate 'difficult' elements such as Pd; (vi) design studies for matching the life of fuel cycle plants with those of reactors to ensure economy; and (vii) development of pyro reprocessing technology for metallic fuel .

## Safety of Fuel Cycle Facilities

Wet processes and inflammable materials are avoided in the fuel fabrication routes to enhance safety. Simplified pelletising, sphere packing and vibro packing are some of the pursued strategies with significant successes. R&D on process control and sensors for process monitoring and early warning systems is required to ensure low probability of major radioactive release to public domain due to any adverse event. Online process control devices such as online alpha monitors (OLAM) prevent letting in high concentration plutonium feed solutions into critically unsafe oxalate supernatant tanks and thus, prevent criticality. The same device detects inadvertent release of alpha emitting plutonium or actinides into potentially active or low active effluent streams.

R&D related to improving corrosion resistance of equipments in reprocessing facilities using aqueous route, has achieved successes. For components using higher concentrations of nitric acid, for example, the dissolver of fast reactor reprocessing plants with 11.5 N nitric acid, titanium and alloys offer good corrosion resistance and corrosion loss is less than 5 mpy (0.13mm/year). To overcome the corrosion of titanium in condensate zones of dissolvers, new alloys like Ti-5%Ta-2%Nb and Zircaloy-4 have been developed, where corrosion loss can be reduced further to 1 mpy (0.025 mm/year).

Remote in-service inspection technologies are necessary to ensure minimum leaks and radioactive contamination. In service inspection techniques like laser triangulation for thickness measurement of dissolver vessel, remote imaging for viewing leakage of waste storage tanks etc. have been developed for application in fast reactor reprocessing plants. On line and In-situ material balance methods for prompt detection of fissile material accumulation are developed and implemented. Neutron collars are systems wherein a number of neutron detectors are placed in an annular cylinder (of moderating material such as high density polyethylene) surrounding the active pipelines carrying high level wastes. The neutron detectors are coupled to increase the sensitivity of the system. In the case of fast reactor fuel cycle facilities, gamma activity is high due to high burn up of fuel and techniques like active neutron interrogation are used for hull and waste monitoring. Development of laser induced breakdown spectroscopy and laser thermal lensing methods are in advanced stage of research for applications related to robust methods for monitoring loss of fissile material in waste streams.

Priority R&D for enhancing safety in aqueous reprocessing are:

- Recovery and burning of minor actinides namely Am, Np and Cm, to reduce radio toxicity of the high level waste and reduce the time period required for reaching low radioactivity levels.

- Partitioning of heat generating nuclides such as Cs-137 and Sr-90.

- Separation of long lived fission products, thereby reducing the radioactivity in wastes.

- Development of human factor models, use of artificial intelligence for early reliable diagnosis, real-time operator aids and automation for short-term accident management.

## Summary

Sodium-cooled fast reactor with closed fuel cycle is an inevitable technology option for providing energy security for India. FBTR has provided substantial and valuable operating, maintenance & safety related experience of a sodium cooled fast reactor. PFBR is a demonstrator and a fore runner in the series of SFRs planned in India. Beyond PFBR, economic competitiveness is important for rapid commercial deployment of SFRs Reprocessing pilot plant, CORAL has given valuable experience in processing a 'difficult' fuel at high burn-up and with short cooling time, which is useful in launching commercial scale reprocessing of PFBR fuel and future oxide fuels. Roadmap with extensive time based comprehensive R&D for large scale deployment of SFRs and systematic introduction of metallic fuel reactors with emphasis on breeding gain and co-located fuel cycle facilities based on Pyro-chemical reprocessing, is being pursued through systematic efforts of the Department of Atomic Energy, collaborations with a large number of academic and research institutions in India and abroad. The successes and planned R&D give the confidence to reach maturity in this challenging technology and attain a leadership position.

# References

B. Raj *et al.* (2009). "Assessment of Compatibility of a system with Fast Reactors with Sustainability Requirements and Paths to its Development", IAEA-CN-176-05-11, International Conference on Fast Reactors and Related Fuel Cycles: Challenges and Opportunities (FR09), Kyoto, Japan.

Chetal, S.C., V. Balasubramanian, P. Chellapandi, P. Mohanakrishnan, P. Puthiyavinayagam, C.P. Pillai, S. Raghupathy, T.K. Shanmugham and C. Sivathanu Pillai (2006). *Nuclear Engineering and Design* 236.

Chetal, S.C. , P. Chellapandi, P. Puthiyavinayagam, S. Raghupathy, V. Balasubramaniyan, P. Selvaraj, P. Mohanakrishnan and B. Raj (2009). "A Perspective on Development of Future FBRs in India", Int. Conference on Fast Reactors and Related Fuel Cycles: Challenges and Opportunities (FR09)', Kyoto, Japan.

Chetal, S.C., P. Chellapandi, P. Mohanakrishnan, C.P. Pillai, P. Puthiyavinayagam, P. Selvaraj, T.K. Shanmugam, C. Sivathanu Pillai (2007). "Safety Design of Prototype Fast Breeder Reactor", ICAPP-2007, France.

Chellapandi, P., P. Puthiyavinayagam, V. Balasubramaniyan, S. Ragupathy, V. Rajanbabu, S.C. Chetal and B. Raj (2010). *Nuclear Engineering International* 240: 2948-2956

Chellapandi, P. , S.C. Chetal and B. Raj (2009). *Pressure Vessels and Piping I, Codes, Standards, Design and Analysis*, eds. B. Raj, B.K. Choudary, K. Velusamy pp 89-102

Gen-IV International Forum Annual Report (2007).

Grover R.B. and S. Chandra (2004). "Strategy for Growth of Electric Energy in India", DAE Publication, Document No-10.

Guidez, J., L. Martin, S.C. Chetal, P. Chellapandi and B. Raj (2008). Nuclear Technology 164.

'INPRO Joint Study on assessment of innovative nuclear systems based on closed fuel cycle with fast reactors', IAEA INPRO-TECDOC-1639

*'Integrated Energy Policy Report of the Expert Committee'* Government of India, Planning Commission, New Delhi (2006)

International Working Group on Fast Reactors, Fourteenth annual meeting, Summary Report, (1981)

International Working Group on Fast Reactors, Meeting on In-service Inspection and Monitoring of LMFBRS, Bensberg, Germany F.R. (1980)

Kakodkar, A. (2007). "Emerging Indian Nuclear Power Programme", Technical Talk at AREVA, Paris, France.

Kakodkar, Anil (2005). Statement and 49th General Conference, Vienna (2005).

Raj, B. *et al.* (2009). "Assessment of Compatibility of a system with Fast Reactors with Sustainability Requirements and Paths to its Development", IAEA-CN-176-05-11, International Conference on Fast Reactors and Related Fuel Cycles: Challenges and Opportunities (FR09), Kyoto, Japan.

Ramachandran, A. and J. Gururaj (1999). "Perspective on Energy R&D and Next Generation Technologies", 86th Session of the Indian Science Congress, Chennai.

Raj, B. (2008). "Science and Technology of Fast Breeder Reactor programme in India: Challenges and Achievements", Prof. Jai Krishna Memorial Award 2008, Annual Convention of INAE, Goa (2008).

*Acknowledgments*: The science and technology contributions by teams of scientists and engineers from IGCAR, BHAVINI, collaborating institutes and Industries, are sincerely acknowledged.

9

C. GANGULY

# An Update of Uranium Fuel Cycle and the Challenges

## India's Access to International Uranium and Fuel Cycle Market

## Introduction

The Atoms for Peace mission was initiated around the middle of 1950s by the United Nations (UN) for ensuring that 'nuclear fission' does not become a tool for human self-destruction but instead a force for prosperity and development of humanity.[1,2] Nuclear fission power reactors were commercially introduced the first time in the world during 1954-1957 in Russia, UK and USA. During the same time, the International Atomic Energy Agency (IAEA) was formed at Vienna, Austria, as an independent technical organisation within the UN family, to facilitate peaceful uses of nuclear science and technology and to inhibit its use for military purpose. Uranium and plutonium are dual use materials that are utilised for peaceful purpose as fuels for nuclear power reactors and also for making nuclear bombs. Hence, prudent management and proliferation-resistance of fissile and fertile materials are of paramount importance in global nuclear commerce and in nuclear fuel cycle activities in order to prevent clandestine diversion of these materials for non-peaceful purpose. In 1970, the IAEA was mandated to the treaty on non-proliferation of nuclear weapons, commonly known as the non-proliferation treaty (NPT), to verify that nuclear material and activities in the non-nuclear-weapons

---

1. D.D. Eisenhower 470th Plenary Meeting of the United Nations General Assembly (1953).

2. Proceedings of the First International Conference on Peaceful Uses of Atomic Energy, Geneva (1955).

States are not used for military purposes. India is not a signatory to the NPT.

The installed nuclear power capacity in the world rose from less than 1 GWe in 1960 to 100 GWe in the late 1970s, and 300 GWe in the late 1980s. Since the late 1980s, worldwide nuclear capacity slowed down mainly because of the Three Mile Island and Chernobyl accidents. However, since the beginning of this century there has been rising expectation for nuclear energy worldwide as a mature, safe and reliable technology capable of generating electricity and supplying heat energy for non-electric application on a large scale, at an affordable price, without pollution or greenhouse gases. Currently, some 435 nuclear power reactors, with installed capacity of ~ 370 GWe, are in operation in 30 countries, generating approximately 14 per cent of global electricity and 62 reactors are under construction. So far, there has been some 15,000 reactor years experience in producing civil nuclear power. Though there is some temporary set-back in the growth of nuclear power following the worldwide recession and the unforeseen disaster at Fukushima Daichi nuclear plant in Japan, in March 2011, caused by unprecedented combined effect of a high intensity earthquake and tsunami, the revised low and high nuclear power projections of IAEA for the year 2030 is still around 501 GWe and 746 GWe respectively.[3]

Nuclear fuels are manufactured from natural uranium (~99.3% $U^{238}$ + ~0.7% $U^{235}$) and natural thorium (~100% $Th^{232}$) resources. Currently, all operating nuclear reactors in the world derive energy from the fission of $U^{235}$, the only fissile material in nature. When $U^{235}$ atoms are bombarded with neutrons, some split into two lighter atoms known as fission products and release enormous amount of heat energy and 2 or 3 neutrons per fission. In most nuclear power reactors, the fission heat energy is utilised for production of steam from water, as shown in Figure 9.1, for generation of electricity. In thermal reactors, the high energy neutrons or fast neutrons released from fission reaction, are slowed down or thermalised by moderators like $H_2O$, heavy water or graphite and the fission chain reaction is sustained mainly by slow or thermal neutrons. In fast reactors, no

3. Power Reactor Information System (PRIS) , International Atomic Energy Agency (IAEA), Vienna (2012).

moderator is used and the fission reaction is initiated and sustained by fast neutrons. The released neutrons not only sustain the fission chain reaction but also transmute some of the 'fertile' $U^{238}$ atoms to fissile $Pu^{239}$ atoms. A series of such neutron capture reactions with $U^{235}$, $U^{238}$ and $Pu^{239}$ in a reactor also lead to the formation of other isotopes of plutonium, namely, $Pu^{240}$, $Pu^{241}$, $Pu^{242}$ and $Pu^{238}$ and Minor Actinides (MAs: Np, Am and Cm) which are either fertile or fissile and do not occur in nature. $Pu^{241}$ is also a fissile isotope of plutonium. Thorium is nearly three times more abundant in nature than uranium, but exists only as $Th^{232}$ isotope, which is 'fertile' rather than 'fissile'. Hence, thorium is not directly usable as a fuel, though it could be transmuted to fissile $U^{233}$ in a reactor. Since global uranium resources are adequate for any foreseeable growth scenario of nuclear power, at least in this century, thorium-based fuels are not likely to be commercially exploited in the coming few decades. However, research and development on thorium fuel and fuel cycle will continue in a limited number of countries, including India, before it attains the industrial maturity of uranium fuel and fuel cycle.

**Figure 9.1**

*Schematic of a Nuclear Power Reactor and Fission and Conversion Reactions*

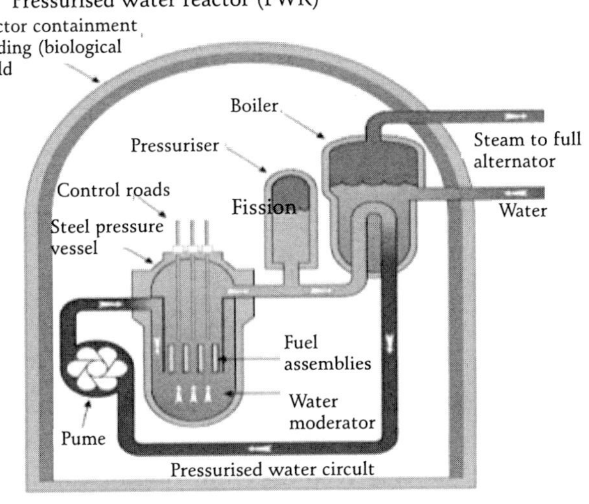


$$\text{Fission} \quad \frac{_{92}U^{235}+_{0}n^{1} \Rightarrow {}_{55}Cs^{140}+_{37}Rb^{93}+3_{0}n^{1}+200\text{MeV}}{_{94}Pu^{239}+_{0}n^{1} \Rightarrow {}_{56}Xe^{142}+_{38}Y^{95}+3_{0}n^{1}+200\text{MeV}}$$

$$\text{Conversion} \quad {}_{92}U^{238}+_{0}n^{1} \Rightarrow {}_{92}U^{239} \Rightarrow {}_{93}Np^{239} \ {}_{94}Pu^{239}$$

Light water cooled and moderated reactors (LWR), consisting of pressurised water reactors (PWR) and its Russian version called VVERs, and boiling water reactors (BWR), account for more than 85 per cent of the operating reactors in the world and are common in USA, Brazil, Belgium, Finland, France, Germany, Spain, Sweden, Switzerland, Russia, Bulgaria, Czech Republic, Hungary, Ukraine, Japan, Republic of Korea (ROK), China and a few other countries. The pressurised heavy water cooled and moderated reactors (PHWR), also known as CANDU, contributes to some 8 per cent of the operating reactors and is the backbone of nuclear power program in Canada and India. A few PHWR units are in operation in Argentina, China, ROK, Pakistan and Romania. Graphite moderated light water cooled reactors, called RBMK, is in operation in Russia in small numbers and is not constructed anymore. The graphite moderated gas cooled reactors namely, Magnox and advanced gas cooled reactors (AGR) are in operation only in UK and are being phased out. BN 600 in Russia is the only commercial fast breeder reactor (FBR) in operation in the world today, but FBR will play a major role in long term sustainability of nuclear power. Two commercial sodium cooled FBRs are under construction, namely, the BN 800 in Russia and the 500 MWe prototype faster breeder reactor (PFBR 500) in India. LWRs will continue to dominate the nuclear power market till the FBRs and related fuel cycle technology attain industrial maturity and become economically competitive.

Since the inception of nuclear power reactors, there has been continuous development and improvement of reactor technology and several generations of nuclear reactors have evolved, as shown in Figure 9.2, with focus on improving safety and efficiency, simplifying design, construction, operation and maintenance and reducing capital cost (Goldberg and Rosner, 2011). Generation I reactors were developed and constructed in 1950 and 1960s but most of these reactors have been retired. The current reactors in operation around the world are mostly of Generation II. The Gen III and III + reactors are relatively new and many are at the stage of construction and planning. These reactors have simple, rugged and standardised designs with lower capital cost, shorter construction time, higher availability and longer operating life (60 years). These reactors have minimum possibility of core melt accident and many have passive safety features. Generation IV reactors are at concept stage and are not likely to be introduced before 2030. The US-led Generation IV International Forum (GIF) aims to develop reactor systems that offer significant advantages in safety, reliability, economics and sustainability.

**Figure 9.2**

*Generations of Nuclear Power Reactors with Progressive Improvements*

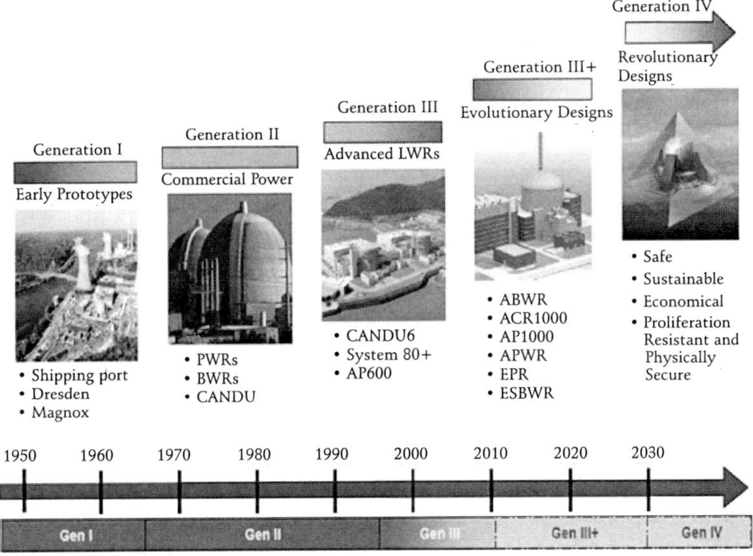


*Source*: Goldberg and Rosner (2011).

The representative fuel bundles and assemblies of operating nuclear power reactors are shown in Figure 9.3. The LWR (including VVER), RBMK and AGR use low enriched uranium (LEU: < 20% $U^{235}$) containing up to 5 per cent $U^{235}$ and the PHWR and Magnox use Natural Uranium(NU) as fuel. Except Magnox, which uses a metallic uranium fuel, all other operating water-cooled reactors and AGRs use high density $UO_2$ pellets as fuel. The fuel pellets are stacked and loaded in zirconium alloy cladding tubes for water cooled reactors and in stainless steel (SS) cladding tubes for AGRs and encapsulated to form fuel pins or rods. The fuel pins or rods are assembled in a particular array with spacers, end plates and other structural components to obtain the fuel bundles or fuel assemblies, which are loaded in reactors.

**Figure 9.3**

*Fuel Assemblies for LWR (PWR/VVER & BWR), PHWR, RBMK and FBR*

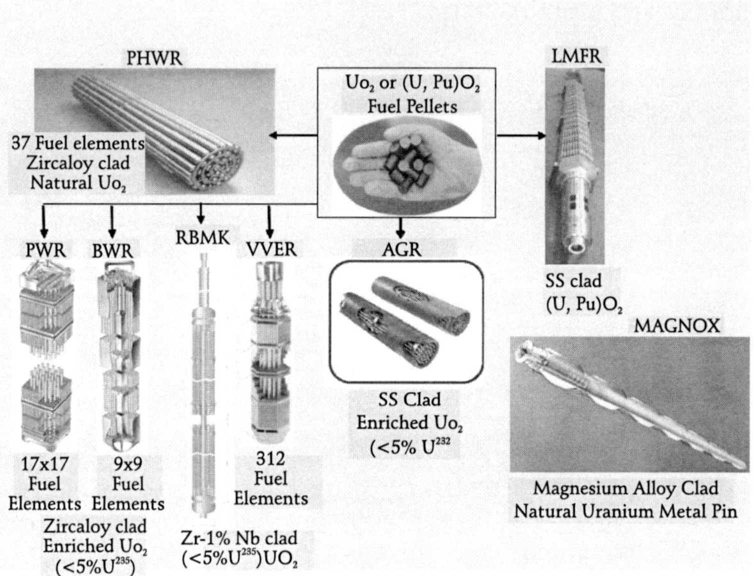


From the inception of nuclear power programme in India in the 1960s, the focus has been self-reliance and indigenisation. Based on the modest uranium but vast thorium resources, a three stage nuclear power program, as shown in Figure 9.4, with 'closed' nuclear fuel cycle was chalked out Department of Atomic Energy (2003). The natural uranium fuelled PHWRs in stage I are linked with FBRs in stage II, working on $U^{238}$-$Pu^{239}$ fuel cycle with depleted uranium (Dep U: $< 0.7\%$ $U^{235}$) or $Th^{232}$ as blanket material. The stage III is made of advanced thermal or fast reactors working on self-sustaining $Th^{232}$-$U^{233}$ fuel cycle. Initially, in the mid-1960s, turn-key contracts were given to General Electric, USA and Atomic Energy of Canada Limited (AECL), Canada to construct two BWR 200 MWe units (de-rated to 160 MWe in 1984) at Tarapur Atomic Power Station (TAPS), Maharashtra and two PHWR 220MWe units (later de-rated to 100 MWe and 200 MWe) at Rajasthan Atomic Power Station (RAPS) at Rawatbhata respectively, mainly for gaining experience in operation and maintenance of nuclear power station and to establish the viability of nuclear electricity in India. These four reactors were put under IAEA safeguards and continue to remain so. TAPS I and II became operational in October 1969 and established the technical and economic viability of nuclear power in India. The performance of TAPS 1 and 2 has remained steady till date and valuable experience has been gained in the operation of BWR. RAPS I started commercial operation in December 1973.

**Figure 9.4**

*Three Stages Indigenous Nuclear Power Programme in India*

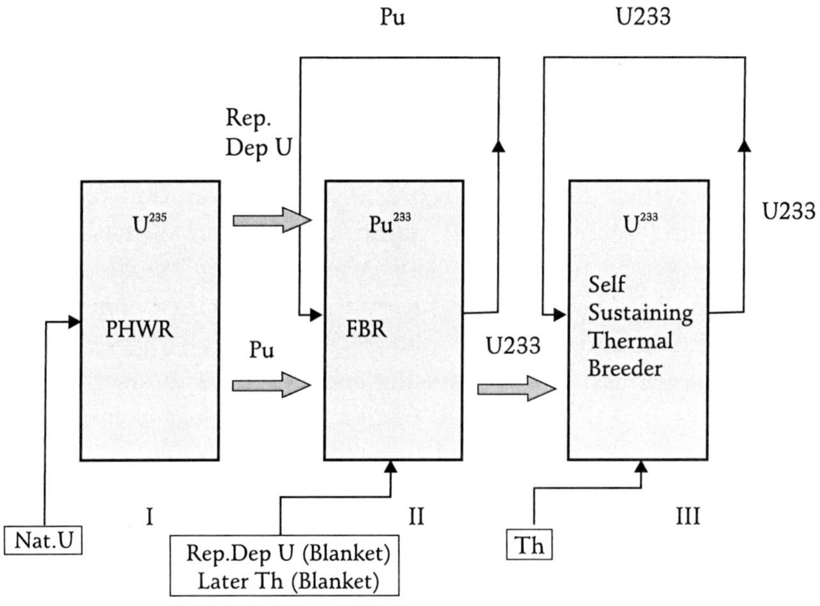


*Source:* Department of Atomic Energy (2003).

India is neither a signatory to the NPT nor did it agree to full scope safeguards of the IAEA and conducted an underground peaceful nuclear explosion (PNE) with a plutonium bomb in the desert at Pokhran, Rajasthan (Pokhran I) in May 1974. The PNE led to India's nuclear isolation from the rest of the world and Canada discontinued the construction activities of RAPS II half way. Indigenous efforts were continued and India could successfully commission RAPS II and put the reactor in commercial operation in April 1981. In subsequent years, India became self-reliant in design, construction, operation and maintenance of PHWR, production of heavy water and manufacturing of nuclear fuel and zirconium alloy core components. Subsequently, India conducted five more underground nuclear explosion tests in May 1998 (Pokhran II) as part of nuclear weapons development program. Today, India has over 350 reactor years of experience from 20 operating small (<300 MWe) and medium (>300MWe but <700 MWe) size reactors with installed capacity of 4780 MWe. Of the 20 reactors constructed so far, 16 are PHWR 220 MWe units, 2 are PHWR 540 units MWe and 2 are BWR 160 MWe units. Over the years, particularly after the post-Pokhran I embargo, the major challenge for rapid expansion of nuclear power in India has been the availability of natural uranium raw material. The uranium resources in India are not only limited but are mostly of very low grades (0.03-0.06% $U^{3O8}$) and occur deep inside the earth.

India's impeccable track record in non–proliferation of nuclear materials and technology, all round cooperation with the IAEA over the years for inspection and verification of its safe guarded reactors and enriched uranium fuel fabrication facility, commitment to restrict $CO_2$ and other greenhouse gas emission that is causing climate change and global warming and rapid economic growth in recent years, paved the way for breaking the nuclear isolation after nearly 34 years . The developed countries could no longer ignore the growing market of nuclear power reactors and the related uranium ore concentrate (UOC) and nuclear fuels in India. In the last quarter of 2008, three consecutive events happened in favour of India. First, the US house of representative passed the US-India Nuclear Cooperation Agreement. Next, the 45-nation Nuclear Supplier's Group (NSG) granted the waiver to India to import civilian nuclear technology, uranium raw material and nuclear fuel. Finally, the IAEA signed an India-specific safeguards agreement. Thus, India is today the only known country with nuclear weapons which is not a party to the NPT but is still allowed to import uranium, nuclear fuel and reactor technology and carry out nuclear commerce with the rest of the world. So far, civil nuclear cooperation agreements have been signed with the USA, Russia, France, UK, Republic of Korea, Canada, Argentina, Kazakhstan, Mongolia and Namibia. Based on these international developments, India has set ambitious targets of installing 20,000 MWe and 62,000 MWe nuclear power by 2020 and 2032 respectively through imported LWRs and indigenous PHWRs and FBRs. Discussions are underway to import, the state-of-the-art Generation III+ LWRs from France (EPR 1650 MWe-PWR), USA (AP 1000 MWe–PWR and ESBWR 1350 MWe–BWR) and Russia (advanced VVER 1000–PWR) with guarantee of life time supply of nuclear fuels for these reactors . The twin units of imported VVER 1000, under construction with Russian collaboration, are in the final stage of commissioning at Kudankulam in Tamil Nadu state, and are likely to be operational in 2012. As part of the construction of indigenous PHWRs, at least 12 units of PHWR 700 MWe have been planned initially of which 4 units are under construction, two each at RAPS and Kakrapara Atomic Power Station.

India has volunteered to put ten PHWR 220 units progressively under IAEA safeguards by 2014 (6 units at RAPS and 2 units each at Kakrapara and Narora atomic power stations) and started importing natural uranium oxide concentrate (UOC) and uranium oxide fuel pellets for the same since 2009. So far, except the 2 units at Narora, the 8 PHWR 220 units are under IAEA safeguards and are using imported uranium for refuelling. During the last three years, UOC from France and Kazakhstan and $UO_2$ pellets from Russia have been imported for further processing at PHWR fuel fabrication plants at Nuclear Fuel Complex (NFC), Hyderabad, that are under IAEA safeguards. Negotiations are underway for importing natural uranium from other major uranium producing countries.

Nuclear reactor and nuclear fuel cycle activities have both reached a high level of industrial maturity in the last four decades in several countries, including India. However, the continuing global recession in the last three years and the unforeseen accident at Fukushima Daiichi plant have underlined the uncertainties confronting the economics and safety of nuclear power and related fuel cycle. Though most countries, with nuclear generation capacity, including India, have reconfirmed their commitment with improved safety and stricter regulatory standards for nuclear reactor and related fuel cycle industries, finding capital for construction of new nuclear power plant and uranium mine is becoming increasingly difficult. There has been drop in uranium spot market price post-

Fukushima and there are no sign of recovery in the last one year. In India, the other impediments have been: (i) the nuclear liability bill passed in the Parliament which is viewed by the vendors abroad and India as overly demanding; (ii) the impasse on administrative arrangements related to uranium procurement from some of the major uranium producers; and (iii) prolonged anti-nuclear demonstration and strikes in several upcoming nuclear power reactor sites.

The present paper summarises the unique features of uranium and plutonium bearing nuclear fuels, the status of supply and demand of uranium ore concentrate and challenges of uranium fuel cycle, highlighting the situation in India, which is poised for rapid expansion of nuclear power and related nuclear fuel cycle activities. The views and recommendation expressed, are the personal opinion of the author, based on his long experience in nuclear fuels and fuel cycle activities and information in open literature.

## Unique Features of Nuclear Fuels

Natural uranium ($U^{235}$ and $U^{238}$) and thorium ($Th^{232}$) and the actinides transmuted from these materials, namely $Pu^{239}$ and other Pu isotopes, MA (Np, Am and Cm) isotopes, $U^{233}$ and $U^{232}$, have the following unique features.

### High Heat Energy Density of Fissile Material and No Greenhouse Gas Emission

One atom of a fissile material, namely, $U^{235}$, $U^{233}$ or $Pu^{239}$, on fission releases nearly 200 MeV energy, which is 50 million times higher than what is generated in the combustion of an atom of carbon in fossil fuel. Thus, 1 kg of natural uranium (containing only ~0.7% $U^{235}$) has the potential to generate 50,000 kWh compared to some 3 kWh and 4 kWh only, produced by 1 kg of coal and 1 kg oil respectively, in conventional thermal power plant. The energy potential of 1 kg $Pu^{239}$, $U^{235}$ or $U^{233}$ is around 60 million kWh. Accordingly, the annual fuel requirement for a typical 1000 MWe nuclear power reactor is only ~30 tonnes as compared to 2.6 million tonnes and 2.0 million tonnes of coal and oil, respectively, needed for conventional thermal power station of equivalent capacity.

Thus, the storage space needed for fresh and spent nuclear fuels are very small. In fact, uranium ore concentrate or finished nuclear fuel for a decade or more could be easily procured in advance and stored in compact storage facilities. Likewise, the annual spent fuel discharged from reactor could be stored for a few decades in the spent fuel storage pool, adjoining the reactor or away from reactor (AFR).

Nuclear fission process does not lead to the emission of $CO_2$, $SO_2$ or $NO_x$, common in thermal power plant, which contributes to global warming, climate change and acid rain.

### Radiation Hazard and the Need for Shielding and Radiation Protection

Natural and man-made isotopes of U, Th, Pu, MA fission products and their daughter products are radioactive and health hazardous to varying degrees and their impact could be genetic and for million years because of the long half-lives of some radio nuclides. Radiological protection of occupational radiation worker and public from the harmful effects of ionising alpha, beta, gamma, X-ray and neutron radiation is of paramount importance to harness nuclear energy. The principle of radiation protection is 'As Low As reasonably Achievable (ALARA)' to minimise the risk of radioactive exposure, while keeping in mind that some exposure may be acceptable in order to further the task at hand. Radiation exposure is universally managed by automation, reducing the time of exposure, remote operation and with proper shielding.

Natural uranium ($U^{235}$ and $U^{238}$) and thorium ($Th^{232}$) are mildly radioactive, emitting mainly alpha particles that can be stopped by a thin piece of paper or even skin. Hence, they have a very little hazard from external radiation and can be handled in the open in well ventilated areas, using protective clothing. Plutonium, minor actinides (Np, Am and Cm), $U^{233}$ and fission products are highly radiotoxic and health hazardous and require proper containment, shielding and remote handling as shown in Figure 9.6. $Pu^{239}$ has a maximum permissible body burden of only 6 x 10-7 g. Hence, any plutonium bearing material has to be handled inside leak tight (alpha tight), well ventilated fume hood, glove box or hot cell maintained at slightly negative pressure. $Pu^{239}$ is essentially an alpha emitter but other isotopes of plutonium emit beta–gamma ($Pu^{241}$) and neutron ($Pu^{238}$, $Pu^{240}$ and $Pu^{242}$) radiation and require additional lead (for beta–gamma) and plastic shielding (for neutron). $Pu^{238}$ and to some extent $Pu^{240}$ and $Pu^{242}$ has, in addition, very high decay heat and require cooling. The isotopes of MAs are also strong beta–gamma and neutron emitters.

**Figure 9.5**

*Plutonium Fuels Fabrication Laboratory*

Plutonium Fuels Laboratory in India             MOX Pilot Plant in Japan

## *Criticality Hazard*

The critical mass is the minimum quantity of fissile material needed, in a particular geometry, to have a self-sustaining nuclear chain reaction. The bare 'critical masses' of 'fissile' isotopes $U^{233}$, $U^{235}$, $Pu^{239}$, $Np^{237}$, $Am^{241}$ in spherical geometry are rather small and in the range of 16 kg, 52 kg, 10 kg, 73 kg and 60 kg, respectively. Hence, only limited and controlled quantity of fissile materials are permitted to be handled, stored or transported at a time to ensure safety from any 'criticality accident'. Precise analysis and real time accounting of fissile material and installation of criticality monitors are essential to ensure 'criticality safety' in nuclear fuel cycle facilities.

## *Generation of 'Fissile' Isotopes from 'Fertile' Ones, while the Former Isotopes Undergo Fission*

Fossil fuelled thermal power stations burn carbon and hydrogen fuel but does not regenerate any carbon or hydrogen in the process. Nuclear fuel is a mixture of fissile (<20 % $U^{235}$ or $Pu^{239}$) and fertile $U^{238}$

isotopes. In a nuclear power reactor, while some of the fissile atoms undergo fission and release fission heat energy, fresh fissile atoms or fuel are simultaneously produced by transmutation of fertile atoms by neutron capture. In water cooled thermal reactors, working on $(U^{235} + U^{238}) - Pu^{239}$ fuel cycle, the ratio of fissile Pu atoms produced to fissile $U^{235}$ atoms consumed, known as 'conversion ratio' is in the range of 0.4-0.6. With multiple recycling of Pu with $U^{238}$ in thermal reactor, a maximum of ~1.0 per cent of natural uranium resource could be utilised. In contrast to FBR working on $U^{238}$-$Pu^{239}$ fuel cycle, the ratio of fissile Pu atoms produced (from fertile $U^{238}$) to fissile Pu atoms consumed for generating fission heat energy, known as 'breeding ratio', is more than 1.0 and could be as high as 1.5 in case of metallic U-Pu fuel. Thus, with multiple recycling of Pu in combination with $U^{238}$ in FBR, it is possible to increase the energy potential of natural uranium resource by a factor of 60 to 70, depending on the efficiency of the fast reactor core design and reprocessing technology.

## Need for 'Proliferation-Resistance' of 'Fissile' and 'Fertile' Material

'Enrichment' of $U^{235}$ in the 'front' end and 'reprocessing' of spent uranium fuel in the 'back' end of uranium fuel cycle, were originally developed in the 1940s for manufacturing $U^{235}$ and $Pu^{239}$ based nuclear bombs and weapons. Thus, enrichment and reprocessing can be used to produce fissile material for nuclear power or for nuclear weapons. The quantity of fissile material needed for producing a nuclear bomb or the bare critical mass, is in the range of 10-60 kg, as mentioned earlier. The basic raw materials needed for producing the fissile materials are natural uranium and thorium. Hence, stringent safeguard procedures including real time accounting of fissile and fertile materials, physical protection of nuclear facilities and intrinsic proliferation—resistant features are required to avoid clandestine diversion of 'fissile' and 'fertile' isotopes for non-peaceful purpose.

## Uranium Fuel Cycle

**Figure 9.6**

*Uranium Fuel Cycle for PHWR and LWR*

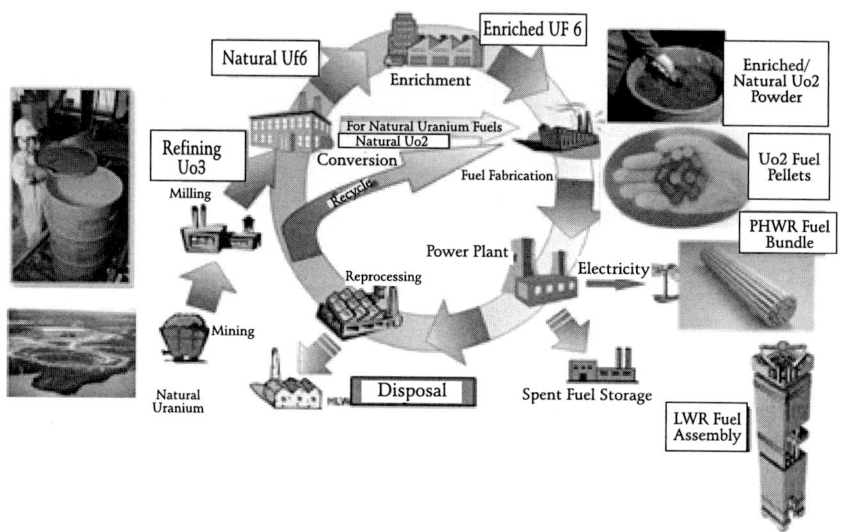


Natural uranium ore, the basic raw material for nuclear fuel, is processed through a series of physical and chemical steps, as shown in Figure 9.6, for preparing uranium fuel assemblies for LWR and fuel bundles for PHWR that are suitable for loading in reactor for generation of fission heat energy. These process steps constitute the front end of fuel cycle. The major steps in front end of nuclear fuel cycle are: (i) mining; (ii) chemical processing or milling of uranium ore to uranium ore concentrate (UOC), popularly known as 'yellow cake' with chemical formula $U_3O_8$; (iii) refining for production of nuclear grade $UO_3$ powder; (iv) conversion to either $UO_2$ for producing natural $UO_2$ fuel powder for PHWR or to $UF_6$ gas for further processing; (v) enrichment of $U^{235}$ in $UF_6$ gas by either gas centrifuge or gas diffusion to the level of 2-5 per cent for LWR, AGR and RBMK; (vi) reconversion of enriched $UF_6$ to enriched $UO_2$ powder; (vii) fabrication of natural or enriched $UO_2$ fuel pellets; and viii) manufacturing zirconium alloy clad $UO_2$ fuel pins/rods and fuel bundles/assemblies for water cooled reactors.

The fuel assemblies remain in the reactor core for some 2 to 5 years depending on the type of reactor and the fuel management scheme after which they are taken out of the reactor and stored in under-water pool, adjoining the reactor. A typical 1000MWe PWR core has some 157 FAs containing ~85 tonnes of $UO_2$ pellets with $U^{235}$ enrichment in the level of ~5 per cent each, fuel assembly has 268 zirconium alloy clad fuel rods in 17x17 square arrays. Each fuel rod is ~4 m long, has an outer diameter of ~10 mm and contains ~500 g enriched $UO_2$ pellets. The reactor is required to be shut down every 12-18 months for refuelling, when some 60 fuel assemblies or one third of the core is replaced. Likewise, a typical PHWR 700 MWe unit has ~5000 fuel bundles containing ~120 tonnes natural $UO_2$ pellets. Each fuel bundle contains ~22 kg natural $UO_2$ pellets, is ~0.5 m long and has 37 zirconium alloy clad fuel pins of ~12 mm outer diameter. PHWRs can be fuelled online. The annual fuel requirement is in the range of 120 tonnes. The natural uranium requirement for a PHWR is nearly 30 per cent lower than that of a LWR of equivalent capacity. In LWRs, a large amount of $U^{238}$ and some $U^{235}$ are locked in tailings of enrichment plant in the form of depleted U.

The spent or used nuclear fuel (SNF) is hot and highly radioactive and is hence, stored under water in a pool for several years, after which the SNF is packed and prepared for permanent disposal in 'open' cycle or subjected to reprocessing, recycling and disposal in 'closed' cycle. In once—through 'open' cycle, the SNF is treated as 'waste' and, regardless of how much fissile and fertile material is left in the SNF, is permanently disposed in underground repository. In 'closed' fuel cycle, the SNF is considered as 'wealth'. After dismantling the SNF assemblies, the fuel is removed from the cladding tubes and subjected to wet or dry reprocessing to separate the Reprocessed U (Rep U) and Pu from the fission products and MA. The Rep U is sent for further treatment for recycling as Uranium Oxide (UOX) fuel in water cooled reactors and the Pu is utilised for fabricating mixed uranium plutonium oxide (MOX) fuel that is recycled in water cooled thermal reactors or preferably in fast breeder reactors. The fission products and MA are vitrified and immobilised in borosilicate glass and prepared for permanent disposal as high active waste in underground repository. In recent years, R & D activities are underway for recovering MA from fission products and recycling the MAs in FBR.

Most of the countries having nuclear power reactors are yet to decide on their fuel cycle option and are following a 'wait and see' strategy. The 'open' fuel cycle option is being followed in Sweden, Finland and USA (may change later). Construction of permanent repositories is underway in Finland and Sweden and is likely to be ready in the 2020s. China, France, India, Japan and Russia have adapted the 'closed' fuel cycle and are pursuing FBR and reprocessing program.

## Uranium Raw Materials: Resources, Production and Demand

Uranium (U) occurs in nature in a range of primary and secondary minerals, of which uraninite ($UO_2$) and pitchblende ($U_3O_8$, $U_3O_7$) are well known. Other common primary and secondary uranium minerals are carnotite (potassium uranium vanadium hydrated mixed oxide), davidite (mixed rare earth ytrrium uranium iron titanate), autunite (hydrated calcium uranium phosphate), torbernite (copper hydrated uranium phosphate), coffinite (hydrated uranium silicates) and uranophane (hydrated calcium uranium silicate).The uranium content in the ore is usually in the range of 0.03 to 0.5 per cent but some deposits contain as high as 20 per cent of U, particularly in Athabasca Basin in Saskatchewan province in Canada.

The world uranium resources are compiled and published, once in two years by OECD Nuclear Energy Agency (NEA) and IAEA. The latest issue, namely the 23rd edition of the Red Book has been published in June 2010 (OECD, 2010). The resources are classified under two main groups, namely the conventional and the unconventional resources. The conventional resources are further divided into 4 categories, namely 'reasonably assured resources' (RAR), 'inferred resources' (IR), 'prognosticated' and 'speculative' resources on the basis of their geological certainty of occurrence. Each of these 4 conventional uranium resources are further grouped in terms of their cost of recovery in US$ per kg of U, namely, <40 US\$, 40 to <80 US\$, 80 to <130 US\$ and 130 to <260 US\$. The combined RAR and IR are known as 'identified resources' of uranium, which is usually used for uranium supply and demand analysis. The undiscovered resources consisting of the prognosticated and speculative categories, are expected to occur based on geological knowledge and regional geological mapping data. Table 9.1 shows the different categories of identified and undiscovered conventional uranium resources and their grouping in terms of recovery cost, based on the information in Red Book.

The total identified uranium resource in the world costing <260 US\$/kg U is 6.3 million tonnes U (OECD, 2010). The total undiscovered resources are 10.7 million tonnes U, including the prognosticated and speculative categories. Australia has the highest uranium resources followed by Kazakhstan, Canada, Russia, USA, Brazil, Namibia, Niger and Uzbekistan. More than 50 per cent uranium resources and production are in countries without any nuclear power as shown in Table 9.2. The uranium produced in these countries are mostly consumed in countries like France, Germany, Spain, Sweden, Switzerland, Japan and Republic of Korea that have no uranium resources or uranium mine and mill. Canada has the highest grade of uranium in the world in their Athabasca basin, where the average U content in Mc Arthur River and Cigar Lake deposits is in the range of 15-20 per cent.

The unconventional uranium resources are speculated to be in the range of 9-22 million tonnes and are mostly in phosphate rocks. Other potential unconventional uranium resources are monazite (thorium ore: mixed thorium uranium rare earth phosphate), marine organic deposits, shale and sea water. Uranium has been successfully produced in the past on a pilot plant scale in USA and Belgium from rock phosphates in Florida and Morocco, respectively. In recent years, some countries including Australia, Brazil, India Jordan, Morocco and USA have initiated R&D activities on recovery of uranium from strong and weak phosphoric acid.[4]

---

4. IAEA Training Meeting/Workshop on Uranium Recovery from Phosphates and Phosphoric Acid, Marrakech, Morocco (2011).

**Table 9.1**

*Categorisation of Identified and Undiscovered Uranium Resources Worldwide*

Decreasing Economic Attractiveness →

Recoverable cost (US\$/kg U)

| | | | | | | *3.0 Mt U Cost unassigned | |
|---|---|---|---|---|---|---|---|
| 260\$/KgU | | 4.0 Mt U | 2.3 Mt U | 6.3 Mt U | 2.90 Mt U | 3.90 Mt U | 6.80 Mt U |
| 130 | several Countries | 3.5 Mt U | 1.9 Mt U | 5.4 Mt U | 2.80 Mt U | 3.70 Mt U | 6.5 Mt U |
| 80 | Australia Kazakhstan Canada | 2.5 Mt U | 1.2 Mt U | 3.7 Mt U | 1.70 Mt U | Nil | 1.70 Mt U |
| 40 | Canada | 0.57 Mt U | 02.3 Mt U | 0.8 Mt U | Nil | Nil | |
| | | Reasonably Assured Resources (RAR) | Inferred Resources (IR) | Total Identified (RAR+IR) | Prognosticated Resources (PR) | Speculative Resources (SR) | Total Undiscovered Resources (PR+SR) |

Identified Resources 6.3 million tons U      Undiscovered Resources 10.7 million tons U

Decreasing Confidence in Estimates

**Table 9.2**

*Uranium Resources and Nuclear Power Reactory: Country-wise Position*

| Country | Resources (Tones U) < 260 US$/kg U (% global resource of ~6.3 million T U) | Production (Tones U) (% global production of 63,085 T U) | No. of Nuclear Power reactors (% total electricity) |
|---|---|---|---|
| *Countries with Major Uranium Resources, Mines and Mills but Without any Nuclear Power Plant* | | | |
| Australia | 1,679,000 (26.62) | 5983 (9.48) | NIL |
| Kazakhstan | 832,000 (13.2) | 19451 (30.83) | |
| Namibia | 284,200 (4.5) | 3258 (5.16) | |
| Niger | 275,500 (4.36) | 4351 (6.89) | |
| Uzbekistan | 114600 (1.82) | 2500 (3.96) | |
| *Countries having Uranium Resources, Mines and Mills and also Nuclear Power Plants* | | | |
| USA | 472,100 (7.48) | 1537 (2.43) | 104 (20%) |
| Canada | 544,700 (8.63) | 9145 (14.49) | 18 (15%) |
| South Africa | 295,600 (4.68) | 582 (0.92) | 2 (5%) |
| Russia | 566,300 (8.97) | 2993 (4.74) | 31 (17%) |
| Brazil | 278,700 (4.42) | 265 (0.42) | 2 (3%) |
| China | 171,400 (2.72) | 885 (1.4) | 11 (2%) |
| Ukraine | 223,600 (3.54) | 890 (1.41) | 15 (47%) |
| India | 1, 05,900 (1.5) | 400 (0.63) | 20 (2%) |
| *Countries with Major Nuclear Power Plants but Without Uranium Resources, Mines or Mills* | | | |
| France | | | 59 (76%) |
| Germany | | | 17 (28%) |
| Japan | | | 53 (25%) |
| Republic of Korea | | NIL | 20 (36%) |
| United Kingdom | | | 19 (13%) |
| Sweden | | | 10 (42%) |

Uranium resources are plentiful and per se do not pose a limiting factor to growth and future of nuclear power .The annual uranium demand during the last few years to fuel installed nuclear power reactors of total capacity ~370 GWe has been in the range of 65,000 to 67,000 tonnes uranium. The projected annual demands of uranium in the year 2035 for low and high growth scenarios of installed nuclear power of 473 and 748 GWe, are 87,370 tonnes and to 138,165 tonnes, respectively (McMurray, 2006). The 6.3 million tonnes identified uranium resources in the world is adequate to fuel the operating and upcoming reactors for at least 100 years on the basis of 'once-through' use in 'open' nuclear fuel cycle, where the uranium utilisation is <1%. The identified uranium will last more than 2500 years in case of FBRs and 'closed' fuel cycle, involving multiple recycling of plutonium and uranium, where the uranium utilisation is 60 times higher than in 'open' cycle. If the entire conventional and unconventional uranium resources are taken into account then the world has enough uranium to fuel nuclear power reactors for several thousand years as summarised in Table 9.3.

**Table 9.3**

*Number of Years of Utilisation of Uranium Resources to Fuel Thermal and Fast Reactors*

| Reactor/Fuel cycle | Number of Years Uranium Resources will last assuming present uranium consumption of ~ 66,500 tonnes/year | | |
| --- | --- | --- | --- |
| | *Using only Identified "U" Resources* | *Using Total Conventional Resources* | *Using Total Conventional and Unconventional Phosphate Resources* |
| Thermal Reactors: "once through open fuel cycle" | 100 | 300 | >675 |
| Fast reactors: "closed fuel cycle and recycling" | >2500 | >8000 | ~20000 |

The annual demand of uranium for manufacturing nuclear fuel for power reactors is mostly met by 'primary supply' or newly mined and processed uranium and to some extent by 'secondary supply', which includes highly enriched uranium(HEU: >20% $U^{235}$) from dismantling of nuclear weapons, civil stockpiles held by utilities and the government, recycled or reprocessed uranium (Rep U) and plutonium from reprocessing plants and re-enrichment of depleted uranium tails from $U^{235}$ enrichment plant. The secondary supply has been meeting some 25 to 30 per cent of the uranium demand during the last few years but these are reducing progressively and are expected to be less than 10 per cent particularly after 2020.[5] HEU of weapons grade (>90% $U^{235}$) has been the main source of secondary supply contributing to 13-15 per cent of global uranium demand. The USA–Russia joint project, titled , 'megaton to megawatt' has been initiated in 1993 to convert 500 tonnes weapon grade HEU, dismantled from some 20,000 nuclear warheads in Russia, to LEU containing ~4.5 per cent $U^{235}$ for use in LWRs in USA. In fact, nearly 10 per cent of electricity in USA is being produced since 1999 with this material and so far some 450 tonnes HEU have been converted (Podvig, 2008; USEC, 2012). The project will come to an end in 2013. Still some 1500 tonnes of HEU and some 260 tonnes of weapon grade plutonium (>93% $Pu^{239}$) are available in the form of nuclear warheads. The USA has offered some 174 tonnes of HEU for conversion to LEU and there has been a USA-Russia agreement to dispose some 34 tonnes weapon grade plutonium for civilian use in the form of MOX fuel in LWRs in USA and FBRs in Russia. It would be worthwhile for India to explore the possibility of having a similar 'megaton to megawatt' project with Russia and USA for importing plutonium from these countries in the form of mixed uranium plutonium oxide (MOX) fuel for the upcoming commercial FBRs after PFBR 500. India should also explore the possibility of importing large quantities of Rep U containing $U^{235}$ in the range of 0.8 to 1.1 per cent for direct recycling as $UO_2$ fuel in the safeguarded PHWRs and extend the fuel burn up.

In India, the Atomic Minerals Directorate of Research and Exploration (AMD) at Hyderabad, under the DAE, is responsible for uranium exploration and resource evaluation. The identified conventional uranium resources is in the range of 105, 900 tonnes, of which nearly 49 per cent are vein type deposits located in Jharkhand state, where mining and milling activities are underway for nearly four decades. The total uranium resources in India stands updated to about 1, 72,400 tonnes of uranium oxide, including some 62,000 tonnes in Tumalapalle, in Cuddapah district, Andhra Pradesh.[6]

---

5. Proc. of IAEA Technical Meeting on Use of Reprocessed Uranium, IAEA-TECDOC-CD-1630 (2007)

6. Government of India (2012). *Annual Report of Department of Atomic Energy, 2011-12.* pp.39-42.

Uranium ore is mostly mined by open pit, underground and In-Situ Leach/Recovery (ISL or ISR) techniques. In open pit and underground mining, the uranium ore is removed from the rock, crushed and ground to fine powder, leached in sulphuric acid or sodium carbonate and bicarbonate and subjected to purification by either Ion-Exchange (IX) or Solvent Extraction (SX), followed by precipitation and drying to obtain Uranium Ore Concentrate (UOC), popularly known as yellow cake ($U_3O_8$). If the uranium ore body is in permeable sand or sandstones, confined above and below by impermeable strata, then ISL, also known as solution mining, can be efficiently utilised. In this technique, the ore is left where it is in the ground and the uranium bearing minerals are recovered by passing sulphuric acid or sodium carbonate and bicarbonate solution through pipes from the surface to the ore body. Thus, the acid or alkali dissolves the uranium bearing minerals preferentially and the pregnant solution is pumped to the surface by another set of pipes and sent to the chemical plant, where the uranium is recovered in much the same way as in any other uranium mill. Thus, in ISL there is a little surface disturbance and no tailings or waste rocks are generated. ISL or ISR mining accounts for ~45 per cent of the total uranium production in 2011 and is followed mainly in Kazakhstan, Uzbekistan, and USA and partly in Australia. Underground and open pit mining contribute to some 38 per cent and 17 per cent of the total uranium production, respectively (World Nuclear Association, 2012).

The world annual production of uranium ore concentrate (UOC) has been in the range of ~53,500 tonnes U during 2010 and 2011. Presently, the annual production of uranium is highest in Kazakhstan (31% of world production), followed by Canada (~14.5%), Australia (9.5%) Niger (7%), Namibia (5%) Russia (~5%) and Uzbekistan (~4%). Table 9.4 summarises the present annual production of major uranium mines worldwide. In 2011, eight leading uranium mining and milling companies in the world accounted for 85 per cent of the world's production of 53,494 tonnes of U. Kazatomprom, with operation only in Kazakhstan, has been the highest uranium producer in the world in 2011 with 8884 tonnes of U, followed by Areva with 8790 tonnes along with operations in Canada, Kazakhstan, Niger and Australia, and Cameco with 8630 tonnes along with operations at the world's largest uranium mine and mill at McArthur River/Key Lake in Canada and ISR/ISL mines and mills in USA and Kazakhstan. ARMZ-Uranium One (7088 tonnes) with mines and mills mainly in Russia and Kazakhstan, Rio Tinto (4061 tonnes) along with operations in Australia and Namibia mainly, BHP Billiton (3353 tonnes) with the Olympic Dam mine and mill in Australia are the other major uranium producing companies.

Uranium ore concentrate, commonly known as 'yellow cake', is traded internationally and is required to be further refined by solvent extraction process to produce nuclear grade $UO_3$ powder. The Blind River uranium refinery in Canada is the world's largest in the world, with an annual capacity of 24,000 tonnes. Nearly 40 per cent of the world's UOC is refined in this facility. In a conversion plant, the $UO_3$ is either reduced to $UO_2$ powder for fabrication of natural $UO_2$ fuel pellets for PHWR, or is converted to $UF_6$ gas for enrichment of $U^{235}$ isotope to the level of 1.5-5 per cent for manufacturing LWR fuel. The tailings in enrichment plant contain $U^{235}$ in the range of 0.15–0.3 per cent and are called depleted uranium (Dep U:< 0.7% $U^{235}$). Dep U is stored as uranium oxide. The enriched $UF_6$ is reconverted to $UO_2$ powder in LWR fuel fabrication plant by dry or wet chemical route.

In India, uranium mining and uranium ore processing or milling are carried out by the Uranium Corporation of India Limited (UCIL), a public sector undertaking (PSU) in the DAE family with headquarters at Jaduguda uranium mine site in Singbhum district of Jharkhand state. The 7 mines and 2 mills in Singhbhum belt produce some 400 tonnes uranium, annually, in the form of magnesium diuranate (MDU) by adapting sulphuric acid leaching followed by ion exchange purification and precipitation as magnesium diuranate. (MDU) (Gupta and Sarangi, 2006). Recently, a new uranium mine and mill has been commissioned at Tummalapalli, Cuddapah district of Andhra Pradesh.[7] In this mill, the uranium ore is crushed, milled and subjected to high pressure alkali leaching

---

7. Tummalapalle Uranium mine to be commissioned in April 2012, *The Hindu Business Line* , Mumbai, February 6 (2012).

**Table 9.4**

*Leading Uranium Mining & Milling Companies in the World in 2011 and their Operation*

| Leading Company | Countries where Uranium Mining & Milling Activities are Underway | | | | | | | | | | Uranium Production in 2011 (tU) |
|---|---|---|---|---|---|---|---|---|---|---|---|
| | Australia | Kazak | Canada | Russia | Malawi | Niger | Namibia | Uzbek | Ukraine | USA | |
| Areva | No | Yes | Yes | No | No | Yes | No | No | No | No | 8790 |
| ARMZ-Uranium One | No | Yes | No | Yes | No | No | No | No | No | No | 7088 |
| BHP Billiton | Yes | No | No | No | No | No | No | No | No | No | 3353 |
| Cameco | Not yet | Yes | Yes | No | No | No | No | No | No | Yes | 8630 |
| KazAtomProm | No | Yes | No | No | No | No | No | No | No | No | 8884 |
| Navoi | No | No | No | No | No | No | No | Yes | No | No | 2500 |
| Paladin | No | No | No | No | Yes | No | No | No | No | No | 2282 |
| Rio Tinto | Yes | No | No | No | No | No | Yes | No | No | No | 4061 |
| VostGOK | No | No | No | No | No | No | No | No | Yes | No | 890 |

with sodium carbonate and the end product is in the form of sodium diuranate (SDU) (Suri, 2010). As on June 2012, the annual production capacity of uranium ore concentrate from eight mines and three mills in India is likely to be in the range of 575 tonnes.[8]

## Procurement of Uranium and Nuclear Fuel: India's Requirements

In conventional thermal power plant, the cost of electricity is extremely sensitive to the cost of fossil fuel, namely, coal, oil or gas and can be rendered uneconomical by even a small increase in fuel price. Nuclear power is almost immune to the price increases of UOC because the cost of UOC is a small fraction ($<5\%$) of the cost of nuclear electricity. Hence, even if the UOC cost is doubled or tripled, the nuclear power cost would only increase marginally.

The utilities with nuclear power plant, procure and purchase natural fuel in a unique way. Instead of buying fuel assemblies from the fabricator, the usual approach is to purchase UOC or yellow cake from uranium mining and milling companies and to buy fuel cycle services like refining, conversion, enrichment, $UO_2$ fuel pellet fabrication and manufacturing fuel bundles or fuel assemblies separately, in order to get the best price and services. Typically a fuel buyer from power plant utility will purchase UOC based on one-time spot market contract or long term multi year contract ranging from 2 to 10 years. Next, they will have separate contracts for fuel cycle services. So far, global trading of uranium had two distinct marketplaces, namely, the western world marketplace comprising mainly of North and South America, Western Europe and Australia and a market place for Russia, Commonwealth of Independent States (CIS), East Europe and China. Most of the fuel requirements for nuclear reactors in the CIS are supplied from the CIS's own stockpiles *http://en.wikipedia.org/wiki/Nuclear_power_plant*. However, in recent years, uranium producers within the CIS have also been supplying uranium and fuel products to the western world and vice versa, thereby increasing the competition.

---

8. Prospects for Increasing India's Uranium Production. *Twentyclicks*.in (2012).

**Figure 9.7**

*Uranium Spot Market Price during the Last 4 Years*

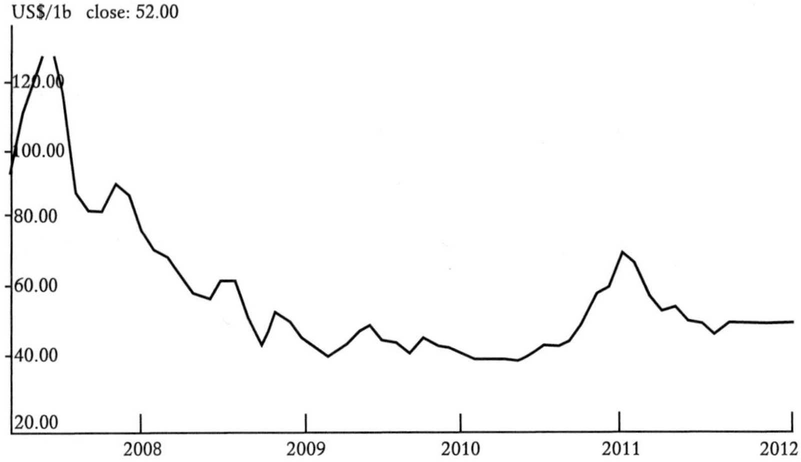

*Source*: Trade tech report on Uranium spot market price (2012).

In general, some 80 per cent to 85 per cent of all uranium is sold under long-term, multi-year contracts ranging from 3 to 10 years, with deliveries starting between 1 to 3 years after the contract is made. Long-term contract terms range from 2 to 10 years, but typically run 3 to 5 years, with the first delivery occurring within 24 months of contract award. They may also include a clause that allows the buyer to vary the size of each delivery within prescribed limits. Figure 9.7 show the variation of uranium spot market price during the last 4 years.[9] From the third quarter of 2007, there has been rapid fall of price, which reached around 40 US$ per pound of $U_3O_8$ in the first quarter of 2010. The market then steadily grew to 70 US$ per pound of $U_3O_8$ in the first quarter of 2011 till the Fukushima Daichi nuclear accident in March 2011. After the accident, the demand of uranium started decreasing mainly because all the reactors in Japan were shut down. Today the uranium spot market price is in the range of 50-52 US$ per pound of $U_3O_8$. The long and medium term price has been in the range of 58 – 62 US$ per pound of $U_3O_8$.

Table 9.5 summarises the uranium requirement for fuelling, operating and upcoming PHWR and LWR units in India, assuming a

---

9. Trade tech report on Uranium spot market price (2012).

plant availability factor of 90 per cent or more (Chauhan, 2009). It is evident that there is a huge gap between the uranium requirement and the indigenous production. This gap has to be filled in through import of uranium from a few reputable companies in the world.

**Table 9.5**

*Natural Uranium (NU) Requirement for the Operating and the Upcoming LWRs and PHWRs in India*

| Reactor type (Mwe) | Avg. Enrichment | EU/unit/ year (Te) | NU/unit/ year (Te) | SWU/unit/year (1000 SWU) |
|---|---|---|---|---|
| TAPS-BWR (160) (2 units) | 2.40 | 9.8 | 46 ( ~ 90 tonnes) | 26 |
| VVER 1000 (6 units) | 3.92 | 23.5 | 188 ( ~ 1110 tonnes) | 134 |
| EPR 1650 (6 units) | 4.95 | 26 | 266 ( ~ 1600 tonnes) | 204 |
| ABWR 1350 (6 units) | 4.20 | 27 | 232 ( ~ 1400 tonnes) | 169 |
| AP1000 (8 units) | 4.80 | 20 | 198 ( ~ 1600 tonnes) | 151 |
| PHWR 220 (16 units) | Nat U : 0.7 | Not Applicable | ~ 45 Tonnes (720 tonnes) | Not Applicable |
| PHWR 540 (2 units) | Nat U : 0.7 | Not Applicable | ~90 Tonnes (180 tonnes) | Not Applicable |
| PHWR 700 (at least 10 Units) | Nat U : 0.7 | Not Applicable | 125 Tonnes (1250 tonnes) | Not Applicable |

## Conventional and Advanced Fuels for LWR, PHWR and FBR

The utilisation of natural uranium resource in thermal neutron reactors like LWR and PHWR is very low and <1%, even after recycling plutonium in these reactors. Most of the uranium is locked as $U^{238}$ in the tailings of uranium enrichment plant, as Dep U in case of LWR fuel and in spent fuel in case of both LWR and PHWR fuels. The spent LWR fuel contains some 3-5% fission products, ~1% residual $U^{235}$, ~1% plutonium, ~0.1% MA and the balance is $U^{238}$, depending on the reactor type, fuel burn up and initial $U^{235}$ enrichment level in fuel. The spent PHWR fuel contains ~0.7% fission products, ~0.2% residual $U^{235}$, ~0.4% Pu and the rest is $U^{238}$. Presently, the spent LWR and PHWR fuels are reprocessed in a few countries to recover U and Pu which could then be recycled in reactor. The fission products and the MAs remain in the raffinate and are fixed by vitrification. The Rep U from spent LWR fuel contains 0.8 to 1.1% $U^{235}$ that can be re-enriched and recycled in LWRs as uranium oxide (UOX) fuel. Alternatively, the Rep U from spent LWR, containing more than 0.7% $U^{235}$, could also be recycled in PHWR to increase the fuel burn up from around 7000 MWD/t HM to around 15,000 MWD/t HM. Direct Utilisation of spent PWR fuel in CANDU (DUPIC) is another alternative, where the spent LWR fuel is not subjected to wet reprocessing for separation of U and Pu from fission products but crushed, milled, compacted into pellets and sintered and recycled in CANDU reactor. Pilot scale fabrication of Direct Use of Pressurised water reactor spent fuel in CANDU (DUPIC) fuel and irradiation testing has been successfully completed at the Korean Atomic Research Institute (KAERI) (Ko and Kim, 2001). The DUPIC fuel cycle has the advantage of inherent proliferation resistance but the economic viability is yet to be established.

Plutonium is best utilised in FBR to breed and burn Pu and also to burn MAs. Mixed uranium plutonium oxide (MOX), containing 15-30 per cent $PuO_2$, is the reference fuel for prototype and commercial SFRs. However, mixed uranium plutonium mono carbide (MC), mono nitride (MN) and U-Pu–Zr alloys are recognised as advanced FBR fuels, mainly on the basis of their higher breeding ratio, higher thermal conductivity and better chemical compatibility with sodium coolant. In the interim period till the FBRs are commercialised, plutonium is being subjected to mono- recycling as MOX fuel, containing up to 10 per cent $PuO_2$ in some 35 LWRs in France, Belgium, Germany, Switzerland and Japan. These reactors can accept up to a maximum of one-third of the core with MOX fuel, the GEN III+ reactors EPR, AP 1000 and ESBWR can accept 100 per cent MOX core. India has successfully demonstrated MOX fuel fabrication and irradiation in the two BWRs at TAPS and in two PHWRs at Kakrapara (Kamath, 2010).

Table 9.6 summarises the conventional and advanced U and Pu bearing fuels for LWR, PHWR and FBR (Ganguly, 1991; IAEA, 2009).

**Table 9.6**

*Conventional and Advanced Nuclear Fuels for LWR, PHWR and FBR*

| Reactors | Conventional Fuels | Advanced/Alternative Fuels |
|---|---|---|
| *LWR (BWR, PWR & VVER)* | | |
| - Fuel | $UO_2$ (U-235<=5%) | - Low enriched UOX ( > 5 % U-235 )<br>- Mixed Uranium Plutonium Oxide (≤12% PuO2)<br>- [LEU+MA] oxide<br>Cr/Cr+Si/ Al/Al+Si dopant as oxide in UOX or MOX for large grain (> 40 $\mu$ ), 9-10 % Gd2O3 + UOX (burnable poison) |
| - Cladding | Zircaloy 2 (BWR)<br>Zircaloy 4 (PWR)<br>Zr - 1% Nb (VVER) | Sn-Nb-Fe & Zr-Nb- O alloys |
| - Burn up (MWd/ton HM) | 20,000-30,000 | High: up to 60,000<br>Ultra High : up to 80,000 |
| *PHWR :* | | - REU or SEU in the form of UO2 |
| Fuel Pellets | Natural UO2 | - (U,Pu)O2 (Th, Pu_O2 & (Th, U233)O2, containing up to 2% fissile material. |
| Cladding | Zircaloy 4 | - PuO2 in Inert Matrix (SiC) Zircaloy 4 |
| Burn Up (MWd/ton HM) | 6,700 MWD/t | 15,000-20,000 MWD/t |
| *Sodium cooled FBR* | He-bonded pins | He-bonded or Na-bonded (U,Pu) C, (U, Pu)N & Na – bonded U-Pu-Zr (≤25% Pu)fuel with /without MA |
| Fuel (pellets/particles/pins) | (HEU)O2 &(U,Pu)O2 (≤ 25% Pu) pellet | - He-bonded vibratory compacted oxide, carbide and nitride pins<br>- 'Pu' and (Pu,MA) in inert matrix |
| Cladding | Stainless Steel D 9 | - Oxide Dispersion Strengthened Steel |
| Burn up (MWD/t HM) | 100,000 | 200,000 |
| Breeding Ratio | 1.0-1.2 | 1.2 – 1.5 |

## Manufacturing of $UO_2$ and MOX Fuels for Thermal Reactors

The manufacturing processes of uranium oxide (UOX) and mixed uranium plutonium oxide (MOX) fuels are more or less similar because $UO_2$ and $PuO_2$ are iso structural (FCC, $CaF_2$ type), completely solid soluble and have similar thermodynamic and thermo physical properties. The two main steps in the fabrication are:

- preparation of oxide in the form of powder, granules or microspheres, starting from suitable compounds of uranium and plutonium;

- consolidation of the fuel powder, granules or microspheres and encapsulation in cladding tubes.

Figure 9.8 shows the major steps and options in preparation of $UO_2$ and MOX powder, microspheres, granules and fuel pins. The industrial methods followed worldwide for preparation of $UO_2$ powder are the Ammonium Di Uranate (ADU) process (Ganguly and Jayaraj, 2004), the Ammonium Uranyl Carbonate (AUC) process (Assmann and Becker, 1979) and the Integrated Dry Route (IDR) (Ion and Watson, 1993). The ADU and AUC are wet chemical processes that use uranyl nitrate or $UF_6$ as starting materials for preparation of both natural and enriched $UO_2$ powder. The IDR use only $UF_6$ as feed material and is the most common process nowadays for manufacturing enriched $UO_2$ powder. The ADU process is mainly followed in Canada and India for preparation of natural $UO_2$ powder suitable for manufacturing very high density $UO_2$ fuel pellets( >96% theoretical density) for CANDU-PHWRs. In ADU process, pure uranyl nitrate solution is reacted with ammonium hydroxide solution to precipitate ammonium diurinate (ADU), which is subjected to controlled air-calcination followed by hydrogen-reduction and stabilisation to obtain sinterable grade slightly hyperstoichiometric $UO_2$ powder. The ADU derived $UO_2$ powder is extremely fine ($<1\mu$m), has specific surface area in the range 2.5-6.0 m2/g and has poor flowability. Additional process steps, namely pre-compaction or roll-compaction, followed by granulation, are needed for producing free-flowing $UO_2$ granules suitable for pelletisation at 300-400 MPa in high speed automatic hydraulic or mechanical presses. The AUC process was developed and followed in Germany and has been extended to Brazil and Republic of Korea. The ex-AUC uranium dioxide powder is coarse (10-20 $\mu$m) and free flowing, has a relatively high specific surface area (5-7 m2/g) and is suitable for direct pelletisation. The process can be extended for co-precipitation of ammonium uranyl plutonyl carbonate (AUPuC), which also leads to free flowing (U, Pu) $O_2$ powder with excellent micro homogeneity of plutonium. The IDR process is widely practised in France, Russia, UK and USA for preparation of enriched $UO_2$ powder, to fabricate LWR and AGR fuel pellets. In IDR process, $UF_6$ gas is subjected to direct pyro-hydrolysis in a mixture of steam and hydrogen to obtain fine $UO_2$ powder, which is granulated and subjected to pelletisation. The specific surface area of IDR derived $UO_2$ powder is relatively low and in the range of 2.0 m2/g, but the stability of the powder is excellent and no passivation treatment is required.

**Figure 9.8**

*Manufacturing Processes and Performance of $UO_2$ Fuel*

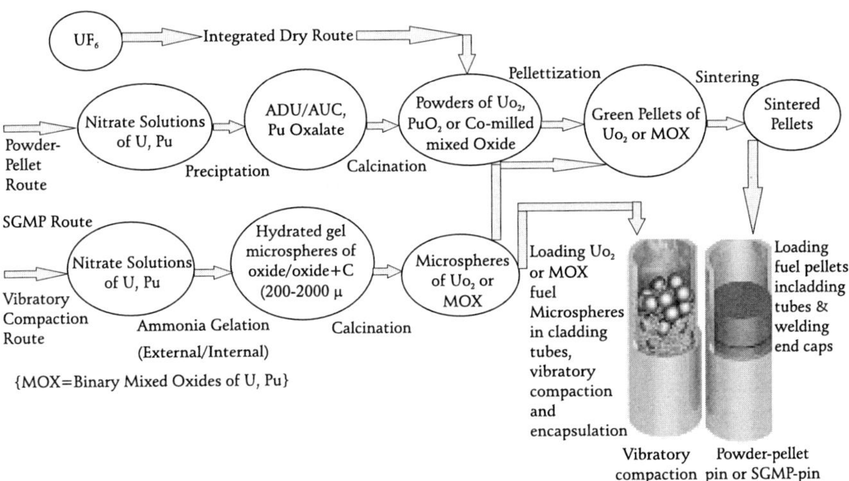


The natural or enriched $UO_2$ powder is subjected to cold-pelletisation to form cylindrical pellets in the diameter range of 10-18 mm, with a length to diameter ratio of 1.0-1.2. The green pellets are sintered at around 1700 C in hydrogen atmosphere, in pusher type or walking beam continuous sintering furnace. The sintered pellets are ground to obtain fuel pellets of desired diameter.

Table 9.7 summarises the current industrial capacity of manufacturing LWR and PHWR fuel worldwide (WNA, 2011). The total capacity of LWR fuel fabrication in atleast 13 countries in the world is around 13000 tonnes, with USA as the largest producer

(3900 tonnes), followed by Japan (~1700 tonnes), Russia (~1400 tonnes) and France (~1400 tonnes). PHWR fuel is manufactured in Argentina, Canada, China, India, Pakistan, ROK and Romania and the total annual capacity is around 4300 tonnes worldwide. Canada has the highest capacity of 2700 tonnes of PHWR fuel, followed by India (600 tonnes) and ROK (400 tonnes).

**Table 9.7**

*Industrial Facilities Worldwide for Manufacturing UO$_2$ Fuel*

| S. No | Country | Natural UO$_2$ (PHWR Fuel) Capacity: tHM/y | Low Enriched UO$_2$ Fuel Assembly (LWR mainly) Capacity: tHM/y |
|---|---|---|---|
| 1. | Argentina | 160 (0.9%U235) | Nil |
| 2. | Belgium | Nil | 700 |
| 3. | Brazil | Nil | 280 |
| 4. | Canada | 2700 | Nil |
| 5. | China | 200 | 450 |
| 6. | France | Nil | 1400 |
| 7. | Germany | Nil | 650 |
| 8. | India | 600 | 25 |
| 9. | Japan | Nil | 1724 |
| 10. | RO Korea | 400 | 600 |
| 11. | Pakistan | 20 | Nil |
| 12. | Russian Federation | Nil | 1400 (includes UO¬2 with HEU for SFR) |
| 13. | Romania | 240 | Nil |
| 14. | Spain | Nil | 300 |
| 15. | Sweden | Nil | 600 |
| 16. | UK | Nil | 860(AGR) |
| 17. | USA | Nil | 3900 |
| 18. | TOTAL | 4320 tHM/y | 12889 tHM/y |
|  | (Rounded): | 4300tHM/y | 13000tHM/y |

The average burn up of PHWR and LWR fuels are in the range of ~7000 MWd/ ton Heavy metal (HM) and 45,000-50,000 MWd/ tHM, respectively. The in-core performance of water cooled reactors has significantly improved over the years. The fuel failures arising out of corrosion and hydriding of zirconium alloy cladding, welding defects of the end caps, and pellet-cladding interaction (PCI) in the event of power ramp are being mitigated by development of improved zirconium alloys with better corrosion resistance and minimum irradiation damage, better quality assurance plan and 100 per cent non-destructive inspection of all end cap welds by advanced ultrasonic and eddy current testing, improved fuel design and fuel microstructure and better in-core fuel management (IAEA, 2010). Efforts are underway to have higher burn up of LWR fuel in the range of 60,000- 70,000 MWd/t HM, with the use of $UO_2$ fuel pellets of large grain size (>40 micron) and having uniformly distributed 'closed' pores in the diameter range of 2-5 micron. This is achieved by admixing dopants like $Al_2O_3$, $SiO_2$, $Cr_2O_3$, $TiO_2$ or $Nb_2O_5$ powder to $UO_2$ powder during co-milling, prior to pelletisation (IAEA, 2004).

*Advanced Method for Manufacturing High Density $UO_2$ Fuel for LWR and PHWR.*

The advanced methods of fabrication of $UO_2$ fuel focus mainly on minimising radioactive aerosol and reduction in operator dose by avoiding generation and handling of fine powder and by remote operation and automation. The other objectives are minimisation of process steps and fabrication cost and improvement in microstructure of fuel for satisfactory operation without failure up to high burn-up. The combination of 'Sol-Gel Microsphere Pelletisation' and 'Low Temperature Oxidative Sintering', known as SGMP-LTS process, developed and deployed at Bhabha Atomic Research Centre (BARC), is an advanced process for manufacturing high density $UO_2$ fuel pellets, avoiding radiotoxic dust hazard associated with fine powder of $UO_2$ and minimising energy requirement during high temperature sintering (Ganguly and Basak, 1991). The SGMP-LTS process has been successfully extended to fabrication of MOX fuel pellets (Ganguly and Basak, 1990). Sol-gel route based on ammonia external and internal gelation process (Ganguly, 1990; Zimmer *et al.*, 1988; Ganguly *et al.*, 1986) is employed for production of hydrated gel-microspheres of uranium oxide in the diameter range of 100 to 1000 micron. Carbon black pore former is added to the sol prior to gelation.

After controlled calcination of the gel-microspheres in air, followed by hydrogen reduction, dust-free, free flowing, porous, soft and coarse $UO_2$ microspheres (10-100 micron) are produced that can be easily pelletised. The green pellets are subjected to low temperature (~1100 C), short duration (~1 hour), oxidative sintering in a 3-zone continuous sintering furnace in nitrogen plus air atmosphere containing around 1000 ppm oxygen to obtain high density $UO_2$ pellets. In LTS, the high cation diffusion coefficient, '$D^U$' of uranium in hyper-stoichiometric $UO_{2+x}$ [$D^U(x) \propto x^2$] is utilised for enhancing the sintering kinetics. For this, 'x' is starting $UO_{2+x}$ and oxygen partial pressures $pO_2$ of sintering furnace in the densification region are closely controlled. After densification, the stoichiometry of $UO_{2+x}$ is adjusted by hydrogen treatment for 0.5 hour at the same temperature in the reduction region of the furnace. $UO_2$ pellets of large grain size recommended to minimise fission gas release can be prepared by adjusting the oxygen partial pressure of the furnace in the densification region and the stoichiometry of the starting $UO_2$ powder. A few 19 pin fuel bundles containing $UO_2$ pellets prepared by SGMP-LTS were successfully irradiated in one of the PHWR 220 unit in Madras Atomic Power Station.

## Challenges in Back End of Nuclear Fuel Cycle

During the last few decades, some 10,500-11,000 tonnes spent fuel were discharged annually from the operating nuclear power reactors in the world and out of the total ~290,000 tonnes discharged, some 90,000 tonnes have been reprocessed, mainly at the La Hague facility in France and Sellafield in UK and to a limited extent in USA, Russia, Japan and India. Most of the industrial spent fuels reprocessing plants follow the 'wet' aqueous route based on Plutonium Uranium Refining by Solvent Extraction (PUREX) process. By the year 2030, some 400, 000 tonnes of used fuel will be generated worldwide (IAEA, 2007). The current world commercial reprocessing capacity is around 5,630 tonnes per year including 1700 tonnes at La Hague, France, 2400 tonnes at Sellafield, UK (1500 tonnes Magnox + 900 tonnes LWR), 800 tonnes at Rokassho, Japan and 400 tonnes at Mayak, Russia (WNA, 2012). However, in recent years reprocessing activities in Sellafield, UK, has been discontinued and commercial operation of Rokkasho plant at Japan has not started and is likely to be delayed indefinitely. Currently, only ~10 per cent of the discharged used fuels are reprocessed and the rest are in interim wet or dry storage facilities. With this trend, long term storage (>100 years) is becoming a reality. Though storage remains an interim solution, both 'open' and 'closed' fuel cycles require geological disposal. So far, no permanent geological repository has been constructed anywhere in the world for permanent disposal of high active nuclear waste. Spent fuel management is, therefore, the major challenge for the future for the nuclear power industry. Public opinion will likely remain sceptical, and nuclear waste disposal will likely remain a topic of controversy, until the first geological repositories are operational and the disposal technologies are fully demonstrated.

Figure 9.9 shows 'closed' uranium fuel cycle combining water cooled thermal reactors in stage 1 with FBR working on multiple recycling of Dep U, Pu and MA in stage 2. Thus, all fissile and fertile materials could be utilised, and high active fission products could be

**Figure 9.9**

*'Closed' U Fuel Cycle Combining Thermal Reactors and Liquid Metal Cooled Fast Reactors (LMFR) Working on Multiple Recycling of Dep/RepU (from PHWR), Pu & MA.*

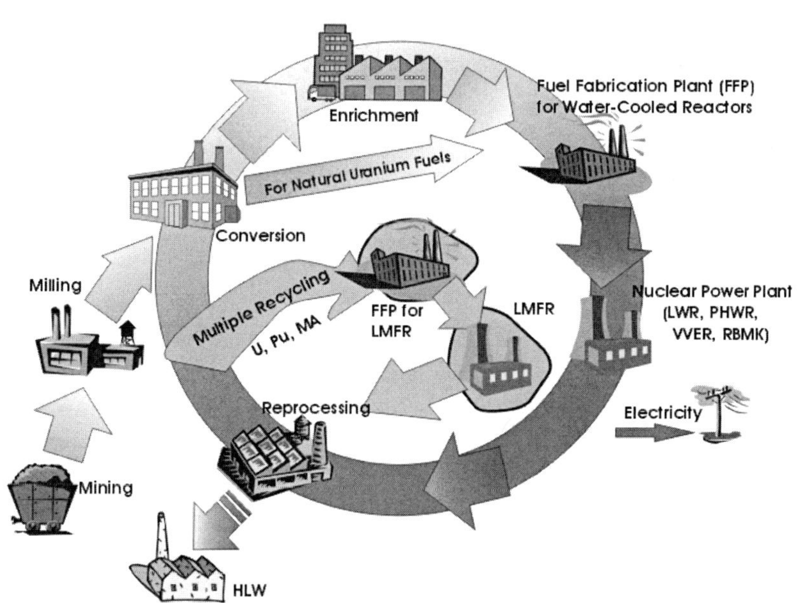

vitrified and disposed permanently in deep and geologically stable repositories.

The Nuclear Energy Agency (NEA) of OECD has recently compiled and published the different 'partitioning' processes, based on 'wet' and 'dry' routes, for reprocessing spent nuclear fuel.[10]

### *Wet Chemical Routes for Reprocessing Spent UOX and MOX Fuel-PUREX Process*

The PUREX process is the most conventional and versatile aqueous or wet chemical route for reprocessing spent uranium metal fuel, UOX and MOX and even mixed uranium plutonium monocarbide and mononitride fuels with some modifications. The PUREX technology has attained a high level of maturity in the last five decades and is being followed on an industrial, semi-industrial, pilot plant and laboratory scales in several countries. Presently, two large facilities, namely the La Hague plants (UP-2 and UP-3 of 1000 tonnes/y each) in France, and the Sellafield plants ( Magnox fuel: 1500 tonnes/year and Thorp 900 tonnes/year) in UK are in operation. 800 tonnes per year reprocessing plant has been set up at Rokkasho-mura, Japan, but commercial operation has not yet started. However, Japan has an operating pilot plant of capacity 90 t/y at Tokai-mura. Reprocessing plants based on PUREX process were also in operation in USA and Germany, but, these have been closed. Semi industrial scale PUREX plants are in operation in India and Russian Federation.

In the PUREX process, the spent UOX or MOX fuel is first dissolved in nitric acid. The nitrate solution, thus obtained is purified by the Solvent Extraction (SX) process, using 30per cent tributyl phosphate (TBP) in kerosene. The uranium and plutonium are extracted to the organic phase, leaving the fission products and MA in aqueous nitric acid phase. Next, the U and Pu are stripped back from organic to aqueous phase by adding nitric acid of very low concentration to obtain pure solution of mixed uranyl plutonium nitrate. The plutonium is separated from uranium by

---

10. Spent Nuclear Fuel Reprocessing Flowsheet- A Report by the Working Party on Fuel Cycle (WPFC) Expert Group on Chemical Partitioning of the Nuclear Energy Agency(NEA) Nuclear Science Committee (NSC), OECD, NEA/NSC/WPFC/ DOC (2012)15 (2012).

redox stripping, in which the oxidation state of the plutonium is lowered by a reducing agent. The pure plutonium nitrate solution, thus obtained, is subjected to oxalic acid treatment followed by air calcination at around 600 C, to obtain $PuO_2$ powder. The oxalate precipitation process is considered to be the standard route for preparation of $PuO_2$. It combines the advantages of high quality product and plutonium decontamination with formation of relatively safe intermediate compounds and solids. The oxalate process also provides some separation between impurities in the nitrate solution and the final oxide product. The uranyl nitrate solution is converted to Rep UOX by the ADU process. The end products from the conventional PUREX process are pure Rep U in the form of uranium oxide, $PuO_2$ powders and high active liquid waste (HLW) containing MA and fission products.

The PUREX technology is being modified in recent years with the objectives of enhancing proliferation resistance of fissile materials by avoiding separate plutonium or other fissile materials stream. The following are some of the few techniques that have been demonstrated so far in pilot plant scale for obtaining MOX and MA-bearing MOX from mixed nitrate solution in modified PUREX process:

- Direct denitration with Microwave Heating (MH) developed in Japan (Suzuki, 2008).

- Co-precipitation of mixed oxalate followed by calcination developed in France (Poinssot and Boullis, 2012).

- Co-precipitation of ammonium uranium plutonium carbonate (AUPuC) followed by calcination developed in Germany (Roepenack *et al.*, 1984).

- Sol-gel 'ammonia external and internal gelation' processes for preparation of dust-free and free-flowing hydrated gel microspheres of mixed oxides in the diameter range of 10 – 1000 micron with or without MA developed in Trans Uranium Institute (Fernandez *et al.*, 2008).

India has over 4 decades of experience with the PUREX process. Currently, the two main semi-commercial reprocessing plants are at Tarapur and Kalpakkam with annual capacity of 100 tonnes spent fuel each (Dey, 2008). These two facilities are not under IAEA safeguards and reprocess only spent PHWR fuel made from natural uranium of Indian origin.

### Dry Pyro-Electrolytic Reprocessing

The so called 'dry' routes of reprocessing are based on high temperature pyro-metallurgy and electro refining techniques to separate actinides from spent nuclear fuel. Molten salts and molten metals are used as solvents. Russia and USA have pioneered the dry route following two different approaches. These methods are now being adapted in Japan, Republic of Korea and China.

### DOVITA Route for Actinide Oxide and Mixed Oxide Fuels

A semi industrial scale pilot plant is in operation for nearly three decades in Russia based on DOVITA (Dry reprocessing of Oxide Fuel, Vibro-pack, Integral, Transmutation of Actinides) for reprocessing spent $UO_2$ or MOX fuel (Bychkov, 1997; Kormilitsyn, 2008). In this route the major steps are:

- chlorination of spent oxide or mixed oxide fuel;

- dissolution in molten salts of NaCl–KCl and NaCl–2CsCl;

- deposition of $UO_2$ and or $(U,Pu)O_2$ with or without MA oxide on the cathode;

- crushing and sieving of the cathode deposit;

- Vibratory compaction of oxide and mixed oxide granulates to form vi-pack fuel pins.

The DOVITA route has been extensively used for manufacturing MOX granules and vibro-packed MOX fuel pins that have been successfully irradiated in BOR 60 and BN 600 FBRs in Russia. The DOVITA process has been extended to preparation of MA bearing experimental mixed oxide pins like $(U, Np)\ O_2$, $(U, Np)\ O_2 - PuO_2$ and $(U, Pu, Am)\ O_2$ containing varying amounts of $NpO_2$ and $AmO_2$.

## Pyroelectrolytic Refining with Molten Salt Electrolyte for Metallic Fuel

The pyro-electrolytic refining of spent metallic U-Zr and U-Pu-Zr fuels from Experimental Breeder Reactor II (EBR II), was first demonstrated in USA in the 1980s. A semi industrial scale plant was set up as part of inherently safe, Integral Fast Reactor (IFR) program of USA, where the FBR, the spent fuel reprocessing plant and the fuel re-fabrication facility are co-located, as shown in Figure 9.10 (Walters, 1999; Till *et al.*, 1997). In IFR, short-cooled, highly radioactive spent fuel is reprocessed by pyro-electro refining process with molten salt electrolyte and separate plutonium stream is avoided by co-deposition with U and, or MA, thereby, enhancing proliferation resistance in fuel cycle. The IFR concept not only ensures proliferation resistance of Pu and other fissile materials but minimises transportation cost of special nuclear material, improves nuclear security and reduces fuel cycle cost.

**Figure 9.10**

*Schematic of Integrated Fast Reactor (IFR) with Metallic Fuel Showing*

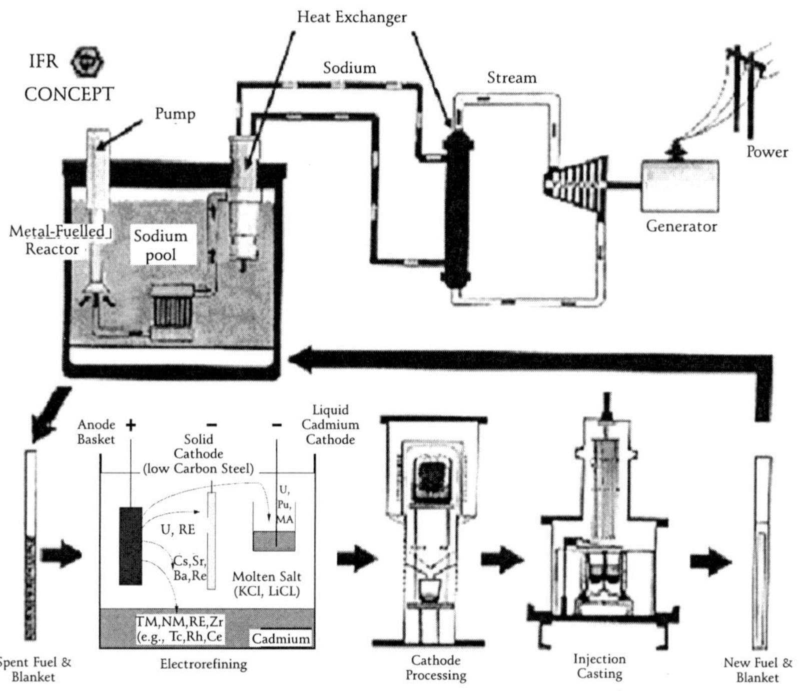


*Source:* Till *et al.*, 1997.

The nuclear fuel cycle includes preparation for electrorefining. The spent fuel pins are chopped into 6-7 mm lengths pieces, loaded into perforated steel baskets for electro-refining and subjected to following treatments:

- addition of $CdCl_2$ to the electrorefining cell at a temperature of 773 K to transfer most of the actinides, sodium and fission products as chlorides to the electrolyte (eutectic mixture of KC1 and LiCl having melting point 350 °C);

- selective dendritic deposition of U on a rotating and cylindrical cathode rod of low-carbon steel (zirconium, molybdenum or uranium may also be used);

- co-deposition of uranium and plutonium on cadmium cathode(liquid cadmium in a beryllia crucible) in the cell as the pre-determined concentration of Pu is reached;

- removal of cathode deposits from the electrorefining cell;

- separation of uranium is from the salt (in case of solid cathode) and U- Trans-Uranium Elements (TRU) from cadmium (in case of molten cadmium) through distillation in a retort;

- remelting U, Pu and MA (optional) for refabrication of U-Pu-Zr alloy fuel pins with or without MA.

*Manufacturing of (U, Pu)O2, (U,Pu)C, (U, Pu)N and U-Pu-Zr Fuels for FBR*

The isotopes of plutonium and minor actinides are highly radiotoxic and many of these isotopes emit neutrons and has high decay heat as shown in Table 9.8. Hence, remote handling and a great degree of automation is needed for manufacturing plutonium and MA bearing ceramic and metallic fuels inside alpha tight glove box and hot cell, with proper beta, gamma and neutron shielding and cooling arrangements.

**Table 9.8**

*Basic Properties of Uranium, Plutonium, Thorium and Minor Actinides Isotopes*

| Isotope | Half-life (y) | Neutron yield (neutrons/sec-kg) | Decay heat (W/kg) |
| --- | --- | --- | --- |
| $^{235}U$ (available in nature) | $700 \times 10^6$ | 0.364 | $6 \times 10^{-5}$ |
| $^{238}U$ (available in nature) | $4.5 \times 10^9$ | 0.11 | $8 \times 10^{-6}$ |
| $^{237}Np$ | $2.1 \times 10^6$ | 0.139 | 0.021 |
| $^{238}Pu$ | 88 | $2.67 \times 10^6$ | 570 |
| $^{239}Pu$ | $24 \times 10^3$ | 21.8 | 2.0 |
| $^{240}Pu$ | $6.54 \times 10^3$ | $1.03 \times 10^6$ | 7.0 |
| $^{241}Pu$ | 14.7 | 49.3 | 6.4 |
| $^{242}Pu$ | $376 \times 10^3$ | $1.73 \times 10^6$ | 0.12 |
| $^{241}Am$ | 433 | 1540 | 115 |
| $^{243}Am$ | $7.38 \times 10^3$ | 900 | 6.4 |
| $^{244}Cm$ | 18.1 | $11 \times 10^9$ | $2.8 \times 10^3$ |
| $^{245}Cm$ | $8.5 \times 10^3$ | $147 \times 10^3$ | 5.7 |
| $^{246}Cm$ | $4.7 \times 10^3$ | $9 \times 10^9$ | 10 |

The manufacturing routes of these fuels must be amenable to automation and remotisation and avoid handling of fine powder to minimise radiotoxic dust hazard.

Mixed Uranium Plutonium Oxide (MOX) , containing 20-30 per cent $PuO_2$ is the universal reference fuel for first generation FBR based on chemical and irradiation stability, high melting point and satisfactory manufacturing and in-pile experience of $UO_2$ and MOX fuels in thermal reactors for more than three decades. However, the breeding ratio of MOX fuel is lower and in turn the doubling time is longer (>25 years) and the thermal conductivity is low. In addition, MOX fuel with higher $PuO_2$ content has the possibility of chemical interaction with sodium coolant if the oxygen to metal ratio is more than 2.00. Mixed uranium plutonium monocarbide (MC), mononitride (MN) and U-Pu-Zr are considered advanced fuels for FBR mainly because of their high breeding ratio and in turn short doubling time (<10 years possible), high thermal conductivity and excellent chemical compatibility with sodium coolant. A significant amount of R&D has been carried out on advanced FBR fuels and these have been very well documented (STI, 2012; IAEA, 2003; Leary and Kittle, 1977; Matzke, 1986; IAEA, 2003).

## Manufacturing Experience of MOX Fuel

Presently, two main processes are followed in the world for manufacturing mixed uranium plutonium oxide (MOX) fuel on an industrial scale. The MELOX plant at France with a name plate capacity of 195 tHM per year is the largest MOX plant in the world and follows the Micronized Master Blend (MIMAS) process (Hugelmann and Greneche, 2000). In the MIMAS process, a mixture of $UO_2$, $PuO_2$ and scrap MOX powder is first subjected to micronisation by co-milling in a ball mill to form a fine master mix of oxide powder containing ~30 per cent $PuO_2$. The master mix in then diluted to the desired plutonium concentration in a secondary blender, by blending free-flowing $UO_2$ powder prepared by AUC/modified ADU process with addition of a lubricant and a small quantity of pore forming agent. Scraps can be recycled at both stages in the process line after a specific treatment. The micronisation during ball milling causes intimate contact between $UO_2$ and $PuO_2$ particles and facilitates complete solid solution formation, during the short period (4-8 hours) of high temperature sintering at 1700°C in hydrogen atmosphere. The Short Binderless Route (SBR) is an integrated process developed by British Nuclear Fuel Limited (BNFL) for manufacturing MOX fuel at the MOX fuel Demonstration Facility (MDF) and later at Sellafield MOX Plant (SMP) of annual capacities of 8 MT and 120 MT respectively (Hugelmann and Greneche, 2000a). The SBR is based on co-milling of $UO_2$, $PuO_2$ and scrap powder in an attritor, followed by granulation in a spheroidiser. Currently, both MDF and SMP are under indefinite shut down. Japan is setting up an integrated facility, named, J-MOX, adjacent to the Rokkasho-mura reprocessing plant, combining advanced aqueous reprocessing and simplified pelletising routes, (Mizuno, 2012). J-MOX facility has incorporated several safety and proliferation resistance (PR) features and will be easy to maintain and produce MOX fuel economically. The mixed nitrate solution containing U, Pu and Np will be subjected to direct denitration by microwave heating (MH) followed by tumbling-granulation to obtain free flowing coarse granules suitable for direct compaction of annular pellets. Semi-industrial scale MOX fuel fabrication plants are in operation at Mayak (Zakharkin *et al.*, 2000) in Russia. Most industrial MOX fabrication plants are devoted presently to LWR fuel fabrication containing upto 10 per cent $PuO_2$, though they could be utilised for manufacturing MOX with higher $PuO_2$ contents, in the range of 20-30 per cent for use in FBRs.

The Advanced Fuel Fabrication Facility (AFFF) plant in India was initially utilised for manufacturing MOX fuel for irradiation testing in the BWRs at Tarapur and in the PHWR at Kakrapara. Later, a few plutonium rich MOX fuel assemblies were manufactured in this facility for the present core of Fast Breeder Test Reactor. Currently, the facility is being utilised for manufacturing MOX containing 20-25 per cent $PuO_2$ for the initial core of PFBR 500 (Panakkal and Mohd. Afzal, 2011). The major feature of the process flowsheet is attritor milling, rotary compaction for preparing annular pellets, high temperature sintering in reducing atmosphere in batch furnace and dry centreless grinding of sintered pellets.

The SGMP-LTS process developed for $UO_2$ and MOX containing 4 per cent $PuO_2$, described earlier (Ganguly and Basak, 1991; 1990), has been successfully extended for manufacturing MOX fuel pellets of controlled density and grain size for fast reactor (Ganguly and Hegde, 1993). The SGMP-LTS route is, particularly, suitable for manufacturing MOX fuel for FBR because of the following advantages:

- the dust-free and free-flowing hydrated gel microspheres of the oxide will be amenable to automation and remote processing in glove box or hot cell;

- the $UO_2$ and $PuO_2$ will be uniformly mixed in the gel microspheres; and

- low temperature ($\sim$1100° C), short duration ($\sim$1 hour) sintering in controlled oxidative atmosphere will reduce electricity consumption, significantly, during sintering and the use of Ar + 8 per cent $H_2$ in the reduction zone will avoid the explosion hazard associated with HTS in $H_2$ atmosphere.

*Manufacturing Experience of (U, Pu) C and (U, Pu) N Fuels*

The mixed uranium plutonium monocarbide and mononitride belong to the same family on the basis of their crystal structure (F.C.C, NaCl type) and similar physical and chemical properties.

They are complete solid solubility. The different techniques of their synthesis and consolidation are similar. The fabrication is difficult and more expensive as compared to oxide fuel because the numbers of process steps are more as compared to that of oxide fuel. In addition, MC and MN are highly susceptible to oxidation and hydrolysis and are pyrophoric in powder form. The entire fabrication is, therefore, required to be carried out under inert cover gas ($N_2$, Ar, He etc.), containing minimal amounts of oxygen and moisture (< 20 ppm each). Further, stringent control of carbon and nitrogen contents is needed during the different stages of fabrication in order to avoid the formation of the unwanted metallic (M) phase and for keeping higher carbides ($M_2C_3$ and $MC_2$) and higher nitrides within acceptable limits. The higher nitrides, however, dissociate to MN at elevated temperature ($\geq 1$ 400oC) in inert atmosphere and pose no problem. The fabrication of (U,Pu) C and (U, Pu)N fuels have, so far, been carried out in a very limited number of countries either on a small laboratory scale or in a pilot plant for small experimental SFRs or for preparing samples for out-of-pile property evaluation and in-pile testing. In depth studies on MC and MN fuels have been carried out in USA, Europe , Japan and India and these have been well documented (STI, 2012; IAEA, 2003; Leary and Kittle, 1977; Matzke, 1986; and Storms, 1967; IAEA, 1974).

The fast breeder test reactor (FBTR) in India is the only reactor using the hitherto untried plutonium rich mixed uranium plutonium monocarbide driver fuel of composition (Pu 0.7 U0.3) C and (Pu 0.55, U 0.45)C that was developed and manufactured in BARC (Ganguly et al., 1986; Ganguly *et al.*, 1994; Ganguly and Roy, 1981; Ganguly *et al.*, 1988). Figure 9.11 shows the different process flowsheets that have been developed for manufacturing (U,Pu)C and (U,Pu)N fuels using $UO_2$ and $PuO_2$ powders or mixed uranium plutonium nitrate siolution as feed materials (Ganguly *et al.*, 1989; Jain and Ganguly, 1993; Ganguly *et al.*, 1991). The two major steps are:

- Carbothermic synthesis of MC and MN using sol-gel derived microspheres containing uniformly distributed carbon particles or tableted mixture of $UO_2$, $PuO_2$ and carbon in vacuum and flowing nitrogen, respectively.

- Crushing, milling, pelletising of MC and MN clinkers followed by sintering in inert or slightly reducing atmosphere to form MC and MN pellets; the sol-gel derived MC and MN microspheres could be directly pelletised and sintered by SGMP process (Ganguly and Hegde, 1997) or subjected to vibratory compaction to form 'vibro-pack fuel pins' (Stratton, 1999).

## Manufacturing Experience of Metallic Fuel for FBR

For metallic U-Zr and U- Pu-Zr fuels, with or without MA, vacuum induction melting, followed by injection casting has proven to be the best technique for remote and automated fabrication of recycled and highly radiotoxic metallic fuel for FBR. In metallic fuel, there is always sodium bonding in the annular space inside the fuel pins. Around 19 weight per cent Zr is added to increase the solidus temperature of the alloys, thereby, providing adequate safety margin during in-pile operation. A pilot plant was set up in USA for remote fabrication of ~35,000 fuel pins of U-Zr and U-Pu-Zr and a few fuel pins of U-Pu-Zr-Np-Am for irradiation testing in EBR-II reactor. Figure 9.11, 9.12 and 9.13 shows the flow sheet for preparation of U-Pu-Zr with or without MA using U, Pu (and MA) and Zr as feed materials (Hofmann *et al.*, 1997; Walters, 1999). First, the feed materials are loaded in a yttria coated graphite crucible and subjected to melting in an induction furnace with dual frequency. At high frequency the field couples with the graphite crucible for heating of the melt while at low frequency the field couples with the melt for a stirring effect, which facilitates homogeneous distribution of U, Pu and Zr in the melt. The yttria coating prevents the melt from reacting with the graphite crucible. The crucibles are capable of repeated use, the melt is heated to about 16000C under an argon atmosphere. The furnace is then evacuated and a pallet containing some 100 one-end closed quartz ($SiO_2$) moulds is immersed with the open end of the molds in the melt. The furnace is immediately pressurised with argon to fill the quartz moulds. The pallet that contains the moulds is lifted from the melt where the cast fuel immediately solidifies and is recovered by crushing the quartz moulds.

**Figure 9.11**

*Process Flowsheet for Fabrication of (U, Pu)C and (U, Pu)N Fuel Pellets*

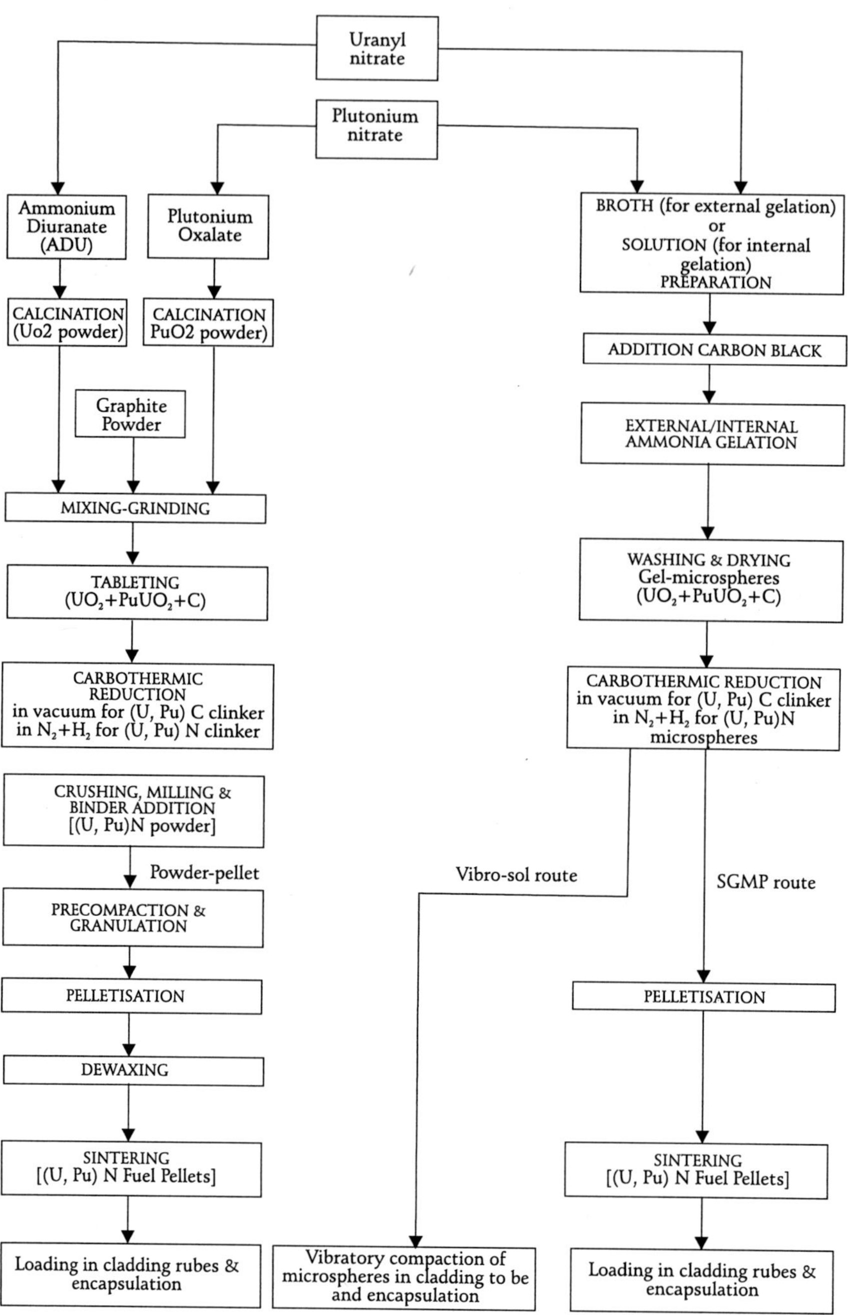

**Figure 9.12**

*Flow Sheet for Fabrication of Sodium-bonded U-Zr and U-Pu-Zr Fuel Pins*

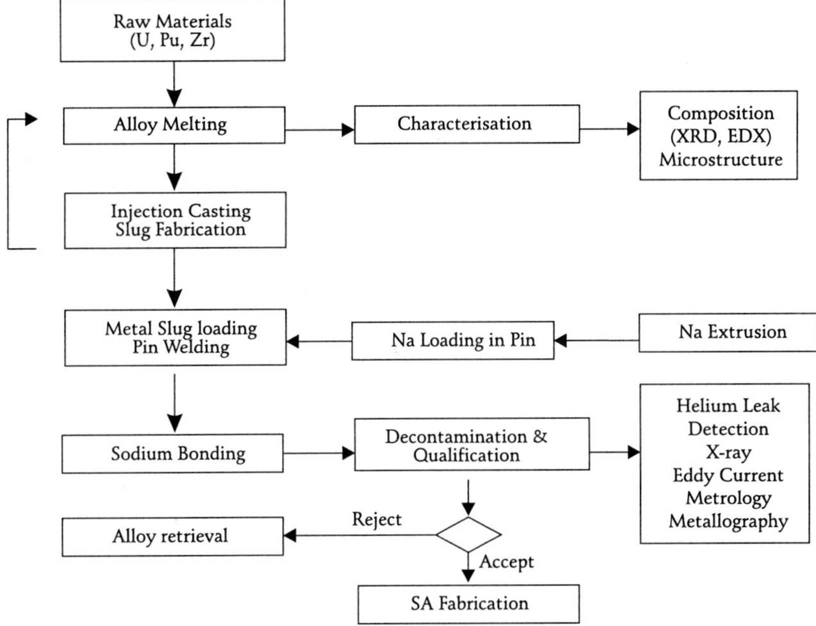


## Possibility of Importing Plutonium and Setting up an International Reprocessing Centre

India is planning to import large numbers of LWRs with life time guarantee for supply of fuel and reprocessing rights and has also started to use imported natural uranium in some PHWR 220 units that are under IAEA safeguards. Hence, there is a need to have at least one centralised reprocessing plant in India, under IAEA safeguards, of much higher capacity, similar to the ones at La Hague, Sellafield or Rokkasho. This safeguarded plant could be an international centre for providing reprocessing service to other countries. The Rep. U from spent fuel of imported LWRs could be utilised as fuel in PHWRs that are under IAEA safeguards. The plutonium from both LWR and PHWR could be recycled in FBR in combination with depleted uranium tailing from $U^{235}$ enrichment plant or reprocessed depleted uranium from PHWR. The proposed 3 stage nuclear power program, involving safeguarded LWR, PHWR and FBR, is shown in Figure 9.13.

**Figure 9.13**

*Proposed 3 Stage Nuclear Power Programmes in India Involving
Safeguarded PHWR, LWR and FBR using Imported Natural U,
LEU and Pu Respectively as fuel.*

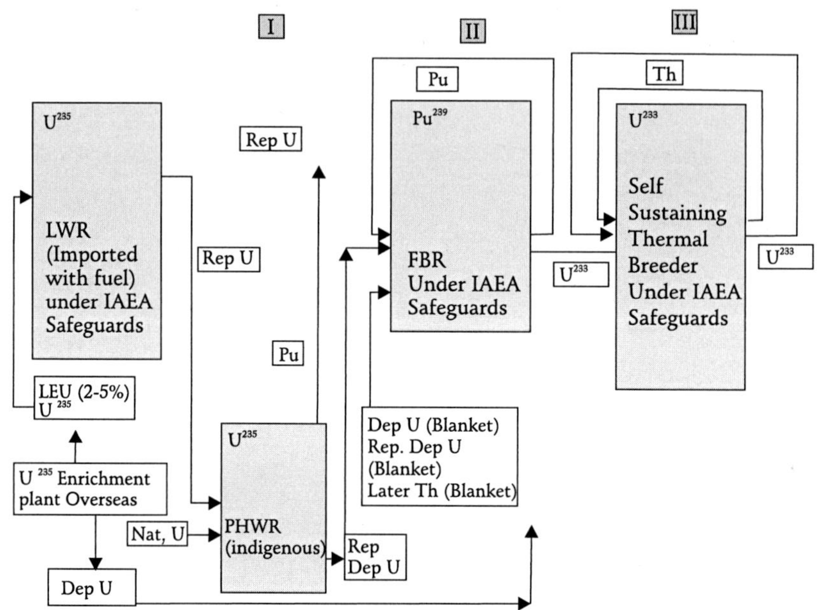


As a follow-up to the PFBR, which is in the final stage of
construction, India is planning to construct six units of 500 MW
(e) capacity based on MOX fuel (Chetal *et al.*, 2012). For this,
availability of plutonium could be a constraint and there could
be a need to import plutonium. The global stockpile of separated
plutonium is estimated as 495 ± 10 tonnes, including some
255 tonnes of civilian plutonium (International Panel on Fissile
Materials, 2012). Russia and the United States have the largest
stockpiles of plutonium produced for weapons. France and UK have
also accumulated large stockpiles of civilian plutonium. India could
explore the possibility of initiating bilateral or multilateral 'megaton
to megawatt' collaboration, with USA, Russia, UK or France and
import plutonium in the form of mixed uranium plutonium oxide
fuel. Around 10 per cent of the global plutonium would be adequate
for the initial and 6 reload fuel cores for at least 4 safeguarded FBRs
of 500 MWe each, after which these reactors will be self-sustaining
on their own.

## Summary and Concluding Remarks

- Nuclear power is one of the attractive options to meet the ever increasing demand of electricity in the world, particularly in India, at an affordable price, in a safe, secured and sustainable manner without causing global warming and degrading the environment. LWR account for more than 85 per cent of operating nuclear power reactors in the world and will continue to dominate the nuclear power market in this century. PHWR contribute to some 8 per cent of the operating reactors and is the backbone of the indigenous nuclear power program in India. Presently, India has 18 small (<300 MWe) and 2 medium ( >300 but <700 MWe) size nuclear power reactors with installed capacity of 4,780 MWe. Hereafter, the upcoming water cooled reactors in India will all be of large size, 700 MWe and more. India has decided to import a number of Generation III+ LWRs of capacity between 1000-1650 MWe from France, Russia and USA (General Electric and Westinghouse) with life time guarantee for supply of fuel and construct some 12 indigenous PHWR 700 units initially. The target is to install some 20.000 MWe and 62,000 MWe nuclear power by 2020 and 2032, respectively. Managing the construction, operation and maintenance and regulatory aspects of so many different types of reactors, will be a big challenge.

- Natural uranium is the basic raw material for $U^{235}$ and $Pu^{239}$/ $Pu^{241}$ bearing fuels. The LWRs and PHWRs use Low Enriched Uranium (LEU) containing up to 5 per cent $U^{235}$ and natural uranium (~0.7% $U^{235}$), respectively as fuel is in the form of high density uranium oxide pellets. The currently identified global resources of natural uranium of some 6.3 million tonnes at a cost of <260 US\$/kg U, is adequate to meet the nuclear fuel demand for any foreseeable growth scenario of nuclear power in the world for at least 100 years in 'once through open cycle', where the uranium utilisation is <1 per cent. If the Pu and Rep U by-products from the operating thermal reactors, including LWR and PHWR, are subjected to multiple recycling in FBR in 'closed' fuel cycle, then the identified uranium resources will meet the fuel requirement for thousands of years.

- Uranium fuel cycle activities have achieved industrial maturity, during the last four decades, in several countries including India, where uranium mining & milling, refining, conversion & fabrication of natural and enriched $UO_2$ fuel pellets and fuel assemblies are being carried out, since the late 1960s to meet the domestic requirements. However, the uranium reserves in India are very limited and the ores are of very low grades containing only 0.03 to 0.06 per cent $U_3O_8$. The annual natural uranium requirement to fuel the operating and upcoming indigenous PHWRs and imported LWRs in India will progressively increase to the level of some 8000 tonnes by 2025. Around 10 per cent of the demand could be met from indigenous mines and mills and the rest has to be imported. Since, the present spot market, medium and long term prices of uranium raw material are reasonably low (50-60 US$/pound $U_3O_8$), large scale import of uranium ore concentrate and or finished pellets from a few reliable UOC and $UO_2$ fuel producers and stock piling the same, will be economic and ensure uninterrupted supply of fuel to the operating and upcoming reactors. Simultaneously, joint ventures in uranium mining and milling with the government/private companies of major uranium producing countries–like Australia, Kazakhstan, Canada, Namibia, Niger, Russia, USA, Uzbekistan etc. are also recommended, at the earliest, for ensuring long term procurement of UOC at reasonable price.

- The major challenge for expanding nuclear power program in India, will be the back end of the uranium fuel cycle. Since, India is following 'closed' fuel cycle and has planned several commercial FBRs, after the commissioning of PFBR 500, large scale reprocessing of spent fuel and availability of plutonium on time for these reactors will be most crucial. Construction of a large international reprocessing plant, under IAEA safeguards, similar to the La Hague, Sellafield and Rokkasho plants, would be necessary, in India in this decade, to reprocess spent fuel from LWRs and PHWRs that are under IAEA safeguards. The international reprocessing centre could also be utilised to provide reprocessing service to other countries following 'closed' fuel cycle. In the interim period, India could explore the possibility of importing some 40 tonnes of plutonium in the form of MOX fuel from France, Russia, UK and USA for the commercial FBRs. Bilateral or multilateral 'megaton to megawatt' collaboration, similar to the on going USA-Russia collaboration on HEU, could be planned to fuel the initial cores and at least four reload fuels for the six commercial FBRs that are planned after PFBR 500. This will pave the way for a strong foundation of peaceful use of atomic energy and fast reactor technology in the world, which is needed for long term sustainability of nuclear power.

## References

Advanced LMFBR Fuels, Topical meeting Proceedings, Tucson, Arizona, Edited by J Leary and H Kittle, American Nuclear Society (1977).

Assmann, H. and M. Becker (1079). Proc. Of the European Nuclear Conference, Hamburg.

Atomic Energy in India (2003). *A Perspective—Golden Jubilee Year Publication.* Department of Atomic Energy.

Bychkov, A. (1997). Proc Int Conf. On Future Nuclear Systems, GLOBAL' 97, Yokohama, AESJ pp 657.

Chauhan, A. (2009). NPCIL, presentation in USIBC-NEI Commercial Nuclear Mission to India, New Delhi and Mumbai.

Chetal, S.C. *et al.* (2012). Proc.Int. Conf on Fast Reactor and Related Fuel Cycles- Challenges and Opportunities, FR 09, 91.

Development status of metallic, dispersion and non-oxide advanced and alternative fuels for power and research reactors, IAEA-TECDOC-1374, (2003).

Dey, P.K. (2008). Spent fuel treatment options and application-An Indian perspective, IAEA-TECDOC-1587, Spent Fuel Reprocessing Options, Vienna.

Fast Reactors and Related Fuel Cycles- Challenges and Opportunities, FR09, Proc. Int. Conf. Kyoto, Japan, STI/PUB/1444 (2012).

Fernandez, A., J. McGinley and J. Somers (2008). Proc Int Conf ATALANTE2008: Nuclear Fuel Cycle for Sustainable Future, Montpellier, paper no 02-06.

Fissile Material Management Strategies For Sustainable Nuclear Energy, IAEA Proceedings series STI/PUB/1288, Vienna (2007).

Ganguly, C. (1990). Metals, Materials and Processes, Science Publications 1, 4, pp 253-274.

———. (1991). Key Engineering Materials, 56 & 57, Trans. Tech. Publications, Switzerland, 435.

Ganguly, C. and U. Basak (1990). Transactions of Powder Metallurgy Association of India 17: 43-48.

———. (1991). Journal of Nuclear Materials 178: 179-183.

Ganguly, C., E. Zimmer and E. Merz (1986). Transactions of American Nuclear Society 52: 44.

Ganguly, C. and P.V. Hegde (1993). *Transaction of Powder Metallurgy Association of India* 20: 23.

————. (1997). *Journal of Sol-gel Science and Technology, USA*, 9, 285.

Ganguly, C., P.V. Hegde and A.K. Sengupta (1991). *Journal of Nuclear Materials* 178: 234.

Ganguly, C. *et al.* (1986). *Nuclear Technology* 72(59).

Ganguly, C., P.V. Hegde and G.C. Jain (1994). *Nuclear Technology* 105(3): 346.

Ganguly, C., P.V. Hegde, G.C. Jain and U. Basak (1989). *Materials Science Forum*. 48 & 49: 153. Switzerland: Trans Tech. Publications.

Ganguly, C. and P.R. Roy (1981). *Fast Reactor Fuel Cycles*. British Nuclear Energy Society (BNES), London. pp.149-156.

Ganguly, C, G.C. Jam, J.K. Ghosh and P.R. Roy (1988). *Journal of Nuclear Materials* 153: 178.

Ganguly, C. and R.N. Jayaraj (2004). *Characterisation and Quality Control of Nuclear Fuels*. Allied Publishers Pvt. Ltd, New Delhi.

Goldberg, S.M. and R. Rosner (2011). *Nuclear Reactors: Generation to Generation*. American Academy of Arts and Science, University of Chicago, USA.

Gupta, R. and A.K. Sarangi (2006). *Emerging Trend of Uranium Mining: The Indian Scenario*. Proc. of IAEA International symposium on Uranium Production and Raw Materials for the Nuclear Fuel Cycle, IAEA-CSP-27 (2006) p.47.

Hugelmann, D. and D. Greneche (2000). Proc. of IAEA / OECD-NEA Conf. on MOX Fuel Technologies for Medium and Long Term Deployment, Vienna, IAEA-CSP-3/pp102.

————. (2000). *The Microstructure of Unirradiated SBR MOX Fuel*, p.160.

Hofmann, G.L., L.C. Walters and T.H. Bauer (1997). *Metalic Fast Reactor Fuels, Progress in Nuclear Energy* 31 (83).

International Atomic Energy Agency (2003). *Status and Advances in MOX Fuel Technology*, IAEA-TRS-415, IAEA, Vienna.

International Panel on Fissile Materials (2012). *Nuclear Weapon and Fissile Material Stockpiles and Production Global Fissile Material Report 2011. www.fissilematerials.org*

Ion, S.E. and R.H. Watson (1993). *Ceram.Eng.Sci.*14 (11-12): 149-154.

Jain, G.C. and C. Ganguly (1993). *Journal of Nuclear Materials* 207: 169.

Kamath, H.S. (2010). *Science Direct*, Elsevier.

Ko, W. and H.D. Kim (2001). *Journal of Nuclear Science and Technology* 38(3): 757.

Kormilitsyn, M. (2008). *Proc. 10th Information Exchange Meeting on Actinide and Fission Product P & T Mito*, OECD/NEA, Paris.

Matzke, H. (1986). *Science of Advanced LMFBR Fuels*, North- Holland, Amsterdam.

McMurray, J.M. (2006). *Worldwide Uranium Resources and Production Capacity: The Future of the Industry*, Proc. of IAEA International symposium on Uranium Production and Raw Materials for the Nuclear Fuel Cycle, IAEA-CSP-27 (2006) pp 27.

Megatons to Megawatts Program Recycles 450 Metric Tonnes of Weapons-Grade Uranium into Commercial Nuclear Fuel, USEC Inc. Press release, USA (2012).

Mizuno, T. (2012). *Proc. IAEA Int. Conf on Fast Reactor and Related Fuel Cycles - Challenges and Opportunities*, FR09, Kyoto, STI/ PUB/ 1444, Vienna.

OECD ( 2010). *Uranium 2009: Resources, Production and Demand*, A joint OECD Nuclear Energy Agency (NEA) and IAEA Report, NEA No 6891.

Podvig, P. (2008). *The Fallacy of the Megatons to Megawatts Program*, Bulletin of the Atomic Scientists Online.

Proceedings of IAEA Technical Committee Meeting on Advanced Fuel Pellet Materials and Designs for Water Cooled Reactors, Brussels IAEA-TECDOC 1416 (2004).

Poinssot, C. and B. Boullis (2012). *Nuclear Engineering International*.

Roepenack, H., F. Schlemmer and G. Schlosser (1984). *KWU/ALKEM-experience in Thermal Pu-recycling*, IAEA Specialists Meeting on Improved Utilisation of Water Reactor Fuel with Special Emphasis on Extended Burnups and Plutonium Recycling CEN/SCK, Mol, NEACRP-L-275.

Panakkal, J.P. and Mohd. Afzal (2011). *Proc. IAEA Technical Meeting on Fast Reactor*, Obninsk, Russia.

*Proc of IAEA Conference on Fuel and Fuel Elements for Fast Reactors I & II*, IAEA, Vienna (1974).

*Review of Fuel Failures in Water Cooled Reactors*, IAEA Nuclear Energy Series, No. NF-T- 2.1, Vienna (2010).

Status and Trends of Nuclear Technologies, Report of the International Project on Innovative, Nuclear Reactors and Fuel Cycles (INPRO), IAEA-TECDOC-1622 (2009)

Storms, E.K. (1967). *The Refractory Carbides I & II*, Academic press, London.

Stratton, R.W. (1999). *Jl Nuclear Materials* 204 (39).

Suri, A.K. (2010). "Innovative Process Flowsheet for the Recovery of Uranium from Tummalapalle Ore", *BARC News letter* issue no. 317.

Suzuki, M. (2008). *Proceedings of 2008 International Congress on Advances in Nuclear Power Plants (ICAPP '08)*, Anaheim, CA, USA, pp 2036

Till, C.E., Y I Chang and W.I. Hannum (1997). *Progress In Nuclear Energy* 31,3.

Walters, L.C. (1999). *Jl of Nuclear Material* 270 (39).

WNA (2011). *World Nuclear Association Report on Nuclear Fuel Fabrication*. London.

——. (2012). *World Nuclear Association Report on World Uranium Mining*. London.

——. (2012). *World Nuclear Association Report on Processing of Used Nuclear Fuel*. London.

Walters, L.C. (1999). *J. Nucl. Mater.* 270 (39).

Zakharkin, B.S. et al. (2000). *Proc. MOX Fuel Cycle Technologies for Medium and Long Term Deployment* IAEA- CSP-3/P.

Zimmer, E., C. Ganguly, J. Borchardt and H. Langen (1988). *Journal of Nuclear Materials* 152: 169.

# IV

## Legal and Regulatory Issues

## 10

R.B. GROVER

# National Framework for Governance of Nuclear Power

## Introduction

Energy is the engine for growth is an oft repeated statement or a cliché, but it does convey the importance of energy for the growth of a country. India has been pursuing nuclear power for a long time and has established a sound framework for its governance so as to address all issues related to safety, security and non-proliferation. The framework consists of legal instruments and policies. Legal framework arises from national acts, and from international conventions, treaties and agreements to which India is a party. I will first give an idea of legal instruments and then come to policies being pursued for the growth of nuclear power.

The Atomic Energy Act, 1962, is the main legislation for governance of nuclear issues in India. Under this Act, the following rules have been framed:

- Atomic Energy (Radiation Protection) Rules, 2004.

- Atomic Energy (Safe Disposal of Radioactive Wastes) Rules, 1987.

- Atomic Energy (Working of the Mines, Minerals and Handling of Prescribed Substances) Rules, 1984.

- Atomic Energy (Factories) Rules, 1996.

- Atomic Energy (Control of Irradiation of Food) Rules, 1996.

- Atomic Energy (Arbitration Procedure) Rules, 1983.

The related acts are:

- Mines and minerals (Development and Regulation) Act, 1957.

- The Weapons of Mass Destruction and Their delivery Systems (Prohibition of Unlawful Activities) Act, 2005.

- The Civil Liability for Nuclear Damage Act, 2010.

Atomic Energy Regulatory Board has been functioning in India since 1983. To convert its, *de facto*, independence to, *de jure*, independence, Nuclear Safety Regulatory Authority Bill, 2011 has been introduced in the Parliament and is under examination by the relevant Parliamentary Standing Committee.

Several notifications have been issued under various acts:

- Setting up of Atomic Energy Regulatory Board (November 15, 1983).

- Prescribed Equipment (Control of Export) Order, 1995.

- Prescribed Substances, Prescribed Equipment and Technology under Atomic Energy Act, 1962 (January 20, 2006).

- Guidelines for Nuclear Transfers (Exports) (February 1, 2006).

- Special Chemicals, Organisms, Materials, Equipment and Technologies (SCOMET) List (September 7, 2007).

- Guidelines for Implementation of Arrangements for Cooperation Concerning Peaceful Uses of Atomic Energy with Other Countries (July 04, 2010).

Through guidelines for nuclear transfers and various lists, Government of India regulates nuclear trade as a responsible nation.

India is also a party to several international conventions and treaties as follows:

- Agreement on the Privileges and Immunities of the IAEA.

- Convention on the Physical Protection of Nuclear Material and its 2005 amendment.

- Convention on Early Notification of a Nuclear Accident.

- Convention on Assistance in the Case of a Nuclear Accident or Radiological Emergency.

- Convention on Nuclear Safety.

- International Convention for Suppression of Acts of Nuclear Terrorism.

- Convention on Supplementary Compensation for Nuclear Damage: signed, yet to be ratified.

India is participating in the Nuclear Security Summit process and honours its obligations arising out of various UN Security Council resolutions. India is not a signatory to Nuclear Non-proliferation Treaty but India's unique status has now been addressed by the international nuclear cooperation related regime. With the relaxation of the guidelines for civil nuclear commerce by Nuclear Suppliers Group (NSG) in September 2008, and signing of "Agreement between Government of India and the International Atomic energy Agency for the Application of safeguards to Civilian Nuclear Facilities," India is now in a position to import uranium and also set up nuclear reactors in technical cooperation with other countries. India is also in a position to make its own contribution as a responsible supplier to the furthering of the international nuclear renaissance.

The basis of the readjustment of the international nuclear regime was the India-US Joint Statement of 2005, where the United States recognised that "India, as a responsible state with advanced nuclear technology, should acquire the same benefits and advantages as other such states". India, in response, reciprocally, agreed that "it would be ready to assume the same responsibilities and practices and acquire the same benefits and advantages as other leading countries with advanced nuclear technology, such as the United States". Therefore, the international community and India showed considerable creativity in working together to modify the international regime to accommodate India's unique position. This has addressed India's position in the international nuclear domain.

Subsequent to relaxation of NSG guidelines, India first signed agreements of cooperation with France (30.08.2008), Russia (05.12.2008) and the USA (10.10.2008). Negotiations with other countries were then taken up and agreements were signed with Namibia, Argentina, Canada, Kazakhstan and Korea; Memorandums of Understanding were signed with UK and Mongolia and a broad based agreement with Russia. One can expect more such agreements in the future.

Before coming to policies, it is necessary to recapitulate certain facts about energy scenario in India. India is a large country, both in terms of area and in terms of population. According to 2011 census, India's population is 1.21 billion, which accounts for about one-sixth of global population. However, its energy resources are nowhere near one-sixth of global energy resources. Indian economy is growing at an impressive rate, but per capita energy consumption is very low—lower than the world average. At present, a significant percentage of energy requirements in India is met by non-commercial energy sources such as fire wood, crop residue and animal waste. Burning of wood in kitchen stoves is unhealthy and all such energy sources need to be replaced by clean sources of energy (Grover, 2008). From the point of end use, electricity is the cleanest source of energy. Switchover to cleaner sources of energy will also contribute towards increase in electricity demand.

Without availability of energy at affordable prices and in a sustainable manner, it is not possible to sustain economic growth and meet expectation of people in India for an improvement in the standard of living. Only way to ensure that India is able to get the energy needed for growth is to find innovative solutions to meet the energy demand. My colleagues and I have been looking at likely growth scenario in India, how demand for energy is likely to increase in the coming decades, what strategies can be adopted to meet the demand and what role nuclear energy has to play. We developed a scenario for growth in electricity demand for the period up to the middle of this century and estimated that electricity production has to be of the order of 8000 TWh by the middle of the century (Grover and Chandra, 2006). This may be compared with the generation in the year ending 31st March 2011, which was about 900 TWh and that quantifies the increase in electricity generation that has to take place in the next four decades.

So what is the solution or what are the solutions? The solutions have to address all issues related to energy both on the demand side and the supply side. One has to work towards improving energy efficiency of economy, explore all sources of energy supply *viz.*, fossil, hydro, nuclear and renewable, develop technologies for their efficient exploitation and considering India's energy resources, this would require innovative solutions at every stage. Planning Commission in India has come out with an Integrated Energy Policy for India and has advocated a similar approach (Planning Commission, 2006). With regard to the role of nuclear energy in India, I would like to say whatever maximum contribution can be made is welcome. Initiative of the Government of India to open up international civil nuclear trade is a step in the direction of increasing the nuclear installed capacity in the country.

Uranium from global sources is now accessible to India. Is that enough? For this, we have to examine global scenario with regard to the availability of uranium and expected growth in nuclear installed capacity. For uranium resources, we can look at the Red Book (OECD Nuclear Energy Agency and the International Atomic Energy Agency, 2007), which gives estimates in four categories depending upon the confidence level about the resource. The categories are: Reasonably Assured Resources (RAR), Inferred Resources (IR), Prognostic Resources (PR) and Speculative Resources (SR). One may ignore SR as they have low confidence level. For growth in nuclear electricity, we have several studies to look at. There is WNA's "Nuclear Century Outlook"; then there is "Nuclear Energy Outlook" by OECD-NEA; and we also have IAEA estimates about growth. One can estimate global requirements of uranium based on these studies and its availability from red book. While numbers might differ, the message is clear. There is not enough uranium to continue to set up thermal reactors.Therefore, unless ongoing exploration locates significant additional uranium resources or extraction of uranium from sea water becomes economically viable, uranium supplies would come under severe stress by the middle of this century. This underscores the importance of fast breeder-reactors to meet global energy demand (Grover *et al.*, 2008). To ensure continued growth of nuclear installed capacity, we have to talk about breeding plutonium and not burning plutonium. There is also plenty of thorium available in the world including India and one has to eventually shift to thorium and breed uranium-233 in suitably designed reactors.

India's reserves of uranium are very modest, but it has vast reserves of thorium. It is because of this reason that India has always advocated a closed fuel cycle approach. In view of its nuclear fuel resource position, growing energy demand and the desire to meet that demand in a manner that can ensure energy independence, it is a necessity for India to take a lead in developing technologies for reactors and associated fuel cycle facilities to close the fuel cycle. Ongoing construction of a Prototype Fast Breeder Reactor, emphasis on developing reactors for using thorium and plans to increase used nuclear fuel reprocessing capacity are steps in this direction.

At present negotiations are ongoing between Nuclear Power Corporation of India Limited and vendors from several countries for setting up light water reactors and when the installed capacity so set up, crosses about 10 GW, setting up uranium enrichment facility in India will be a wise decision from the point of view of energy security. Therefore, India proposes to reprocess used fuel arising from light water reactors as well.

With this background, I can now articulate policies with regard to nuclear power being pursued by India as follows:

- India is committed to growth of nuclear installed capacity so as to provide energy security in a manner that is environmentally sustainable.

- India has unwaveringly followed closed fuel cycle approach and this will continue to be an integral part of India's nuclear power policy.

## References


Grover, R.B. (2008). *Int. J. Global Energy Issues* 30,1/2/3/4 pp 28-248.

Grover, R.B. and S. Chandra (2006). *Energy Policy* 34: 2834-2847.

Grover, R.B., B. Purniah and S. Chandra (2008). *Atoms for Peace: An International Journal* 2(1): 68-82.

OECD Nuclear Energy Agency and the International Atomic Energy Agency (2007). Uranium: Resources, Production and Demand, A Joint Report ("Red Book").

Planning Commission (2006). *Integrated Energy Policy: Report of the Expert Committee*. New Delhi: Planning Commission, Government of India.

11

R. RAMACHANDRAN

# Nuclear Safety

## *The New Regulatory Framework*

The serious nuclear accident at the Dai-ichi nuclear power station in Fukushima situated along the Pacific coastline of Japan in March 2011, following a massive earthquake in the Pacific and the consequent tsunami of unprecedented severity, has prompted several countries to review their respective nuclear safety regulatory framework. In India, the accident revived the old questions about the independence and autonomy of the existing system (Gopalakrishnan, 1999; Hindu, 2011), which vests the responsibility of regulatory functions with the Atomic Energy Regulatory Board (AERB), a wing of the Department of Atomic Energy (DAE) itself. The AERB is entrusted with the task of ensuring safety of nuclear installations and other facilities dealing with radioactivity in the country.

Faced with renewed criticism of the current regulatory structure and at the same time to ally public fear, the Prime Minister forthwith assured the people of the country that a truly independent nuclear regulatory body would be set up.[1] Accordingly, the government tabled the Nuclear Safety Regulatory Authority (NSRA) Bill in the LokSabha on September 7, 2011, towards the end of the monsoon session of the Parliament[2], which seeks to establish a new regulatory authority that will replace the existing AERB.

The AERB was constituted on November 15, 1983, by an Executive Order (Gazette Notification No. 25/2/83) under Section

---

1. *The Hindu* March 30 (2011).

2. *http://164.100.47.5/newcommittee/press_release/bill/Committee%20on%20S%20and%20 T,%20Env.%20and%20Forests/Nuclear20Safety%2076%20of%202011.pdf*

27 of the Atomic Energy Act (AEA) of 1962, and is administered by the DAE. This particular section of the AEA allows the DAE to establish a suitable authority to discharge its duties related to nuclear safety under Sections 16, 17 and 23 of the Act. The criticism of the present regulatory framework has been on the following grounds.

By the very manner in which it was set up, the AERB has been a subordinate entity of the DAE, and therefore lacking in strict functional independence and autonomy. The AERB reports to the Atomic Energy Commission (AEC) whose chairman is the Secretary of the DAE. This means that the regulatory body reports to the very agency whose nuclear activities and operations it is mandated to monitor and regulate to ensure safety in public interest. The AERB depends on the DAE for its funding. In addition, the AERB itself does not have its own technical staff and facilities to discharge its regulatory functions. In fact, there is hardly any expertise regarding nuclear matters outside the DAE system. Thus, the AERB has to rely on the scientists and engineers of the DAE institutions for technical support and expertise. Thus, the demand for a properly empowered independent body in a new framework has been a longstanding one.

It should be pointed out here that Article 8.2 of the international Convention on Nuclear Safety (CNS) of the International Atomic Energy Agency (IAEA) of 1994 (INFCIRC/449), of which India is a signatory, states: "Each Contracting Party shall take the appropriate steps to ensure an effective separation between the functions of the regulatory body and those of any other body or organisation concerned with the promotion or utilisation of nuclear energy." The existing Indian regulatory structure would seem to contravene this. To rectify this, the AERB constituted a committee in 1996, which produced a report titled: *Code for Governmental Organisation for Regulation of Nuclear and Radiation Facilities*. The report apparently had recommended a 'functionally autonomous body' to carry out the regulatory functions. But this report has, for some reason, not been acted upon. (It may, however, be noted that, while India signed the CNS in 1994, it was ratified only in 2005.)

The DAE's argument against such a move has all along been that the country's nuclear programme is at present a small one and a separate organisation can be considered as the programme expands, evolves and extensive experience is gained. In any case, the DAE has always maintained that there has always been complete independence and functional autonomy to the AERB. This is reiterated in the 'Statement of Objects and Reasons' that accompanies the NSRA Bill. This, of course, the critics of the present structure would reject categorically.

In the wake of the renewed impetus after Fukushima to the longstanding debate of the present status of the AERB as an entity subordinate to the DAE in ensuring nuclear safety, the government, as mentioned in the beginning, moved a bill on September 7, 2011, to put in place a truly independent and autonomous nuclear regulatory body. But, it is not clear if the structure articulated in the new NSRA Bill has anything in common with that 16 year-old report on code for regulating nuclear activities. The Bill was referred to the Parliamentary Standing Committee on S&T on September 16, 2011, whose report on the Bill was placed in the House on March 21, 2012.[3] After the changes recommended by the Committee are incorporated, the Bill will be considered by both the houses, before it is passed and promulgated as the Nuclear Safety Regulatory Authority Act.

As the title of the Bill implies, the government seeks to establish a new Nuclear Regulatory Authority (Section 8 of the Bill). The Authority will include a chairperson, two whole-time members and not more than four part-time members, who will be selected by a search committee (Section 9). The general structure of the NSRA is similar to that of the AERB but, unlike the AERB, the NSRA will not be under any ministry. The total number of members now intended, is more than what the AERB is allowed to have. The original EO that set up the AERB required that the total Board strength should not exceed five. However since 1987, the Secretary has been made a non-member and the Board's strength was five

---

3. *http://164.100.47.5/newcommittee/reports/EnglishCommittees/Committee%20on%20S%20and%20T,%20Env.%20and%20Forests/221.pdf*

excluding the Secretary. After the Fukushima accident, however, the strength has been increased to six as it was felt that that the Board needed the expertise of an Earth Science specialist as well. The Chairperson and members can hold office for three years and are eligible of appointment for another term (Section 11).

> The Secretary, DAE, deposing before the Committee stated that the four part time Members representing different disciplines relevant to the functions of the Authority would contribute their expertise and take equal responsibility in the decision making process. He also stated that though they would be part time Members of the Authority, they would be permanent Faculties/Directors drawn from prominent institutions like IIT, CSIR, etc. and that they will not be salaried employees of the Authority. However, expressing its reservation over the nomenclature 'Part-time Members', the Committee has recommended that they be referred to in a more dignified manner in keeping with the dignity of the Authority. The Committee also noted that that the provision as contained in the Bill did not specify as to how many times the reappointment was permissible. It observed that if the provision were to be used to grant repeated extensions at the discretion of Govt., it may impinge on the functional autonomy of the Authority. It therefore recommended that the Chairperson and the Members of the NSRA shall be eligible for reappointment for not more than one term of three years.[4]

Section 18 of the Bill provides for the dissolution of the AERB and transfer of all its assets and liabilities and nearly all its functions once, the NSRA is established in accordance with Sections 8 and 9. The functions of the Authority will, thus, essentially be the same as that of the AERB. According to Section 20 (1) of the Bill, the functions of the Authority, *within its jurisdiction*, include taking appropriate measures "To ensure that the use of radiation and atomic energy is safe for the health of the radiation workers, members of the public and the environment." Clauses 20 (2) (a-s) elaborate on these functions and their implementation.

---

4. Here and in what follows in the text of the presentation, observations and changes recommended by the Committee will be indicated as sub-text in squared parentheses and in smaller font.

Further, to clarify matters, an explanatory note to Section 20 (1) says: "For the removal of doubts...the functions of the Authority shall be confined only to ensure radiation safety and nuclear safety during activities relating to production, storage, disposal, transport, transfer by sale or otherwise, import, export and use of any material, radioactive material or any other substance or equipment, and physical security of nuclear material, radioactive material and radiation and nuclear facilities, and in no case shall extend to functions or any other matter which the central government is required to discharge under the AEA, 1962." According to Section 28, written consent of the NSRA is required for carrying out activities such as production, storage, disposal, transport, export and import of nuclear material or equipment for production or use of nuclear material.

However, Section 20 (3) allows the Authority to exempt from its purview through appropriate notification, "for reasons to be recorded in writing,...any radioactive material, any class or classes of radioactive material or any radiation generating plant from the applicability of any of the provisions of the safety related regulations or orders issued under this Act." This exemption provision has resonance with provisions under Section 25 and 27 of the Bill, which will be presently discussed, which will also clarify the italicised phrase 'within its jurisdiction' above. (p.248) The reasons for the other italicised word 'nearly' (p.248) will also be shortly clarified.

Section 3 of the Civil Liability for Nuclear Damage Act 2010 requires the AERB to notify the occurrence of a nuclear incident within 15 days. Now, the NSRA will assume the same function and Section 20 of Bill requires the Authority to apprise the National Disaster Management Authority (NDMA) regarding nuclear safety and radiation safety measures and management of disaster arising from notified nuclear incident and coordinate with the NDMA, in case of such disaster. The Liability Act has been suitably modified reflecting this change in the notifying authority.

Section 19 of the Bill states: "Save as otherwise provided in Sections 25 and 27, the jurisdiction of the Authority shall extend to all areas to which this Act is applicable and activities relating to production, development or use of atomic energy and radiation in all its applications, or transport (within India or outside India), transfer by sale or otherwise, import, export or storage or disposal of nuclear and radioactive material." Sections 25 and 27 exempt certain areas from application of the provisions of this Bill for reasons of national defence and security. This would include activities and areas related to the nuclear weapons and other strategic programmes involving the DAE, the Defence Research and Development Organisation (DRDO) and the military. Section 25 also provides for establishing 'one or more regulatory bodies' through the government orders, to carry out safety functions in these exempted areas.

> Noting that Section 25(2) only stipulates "The Central government may...establish one or more regulatory bodies..." the Standing Committee report observed that the word 'may' leaves scope of discretion to establish or not to establish such additional regulatory bodies for ensuring nuclear safety in NSRA exempt areas. The Committee felt that such a discretion was not justified as threat to public safety from nuclear material and facilities exempted from the purview of the NSRA was as serious as from the ones within its purview and as such there was an equal sense of urgency and necessity to regulate them too. The Committee, therefore, recommended it should be mandatory for the government to bring the facilities and activities exempted under Section 25(1) of this Bill under one or other regulatory body.

While Section 25 exempts "any nuclear material, radioactive material, facilities, premises and activities" and associated assets and areas from NSRA's jurisdiction for reasons of national security, Section 27 envisages possible expansion of such areas and allows the government to exempt, by order, any new area, nuclear material, radioactive material, nuclear facility or plant from the jurisdiction of the NSRA. Section 27 actually goes much further than that. It envisages the possibility of such activities and areas without the oversight of any constituted regulatory body. It says: "The Central Government may, for the purposes of national defence and security, by order, exempt any area, nuclear material, radioactive material, nuclear facility or plant from jurisdiction of the Authority or other regulatory bodies under this Act and carry out those functions itself in relation thereto, which, but for such exemption, would have been carried out by the Authority or other regulatory bodies under this Act."

It may be recalled that in 2000 the Bhabha Atomic Research Centre (BARC), the key centre of India's weapon's programme, was removed from the AERB's purview through an office memorandum of April 25, 2000 (Ramachandran, 2000). Such exempted activities/areas existed even earlier, following Pokhran-I (1974), but, being not as widely distributed as perhaps it is after Pokhran-II (1998), they were being kept outside AERB's jurisdiction by informal arrangements. Now with the separation of regulatory activities from the DAE, the NSRA Bill incorporates a formal arrangement that demarcates regulatory activities of civilian and military programmes as it is in other Nuclear Weapon States (NWSs).

> It is interesting is to note, however, that the Standing Committee did not think it necessary to change Section 27, which envisages activities and areas outside the purview of not only the NSRA but also other regulatory bodies that may be established for safety of areas exempted under Section 25.

The Bill also provides for establishing an overseeing body called the Council of Nuclear Safety (Section 5). "The Council," says the Section 7 of the Bill, "shall oversee and review the policies with respect to radiation safety, nuclear safety and other matters connected therewith or incidental thereto..." In the current scheme of things, it is the AERB, which is responsible for developing safety policies keeping in view recommendations and evolving standards of international bodies as well as local requirements. As per the proposed new framework, it is the Council, and not the NSRA, which will have the responsibility of evolving nuclear and radiation safety policies. (This explains, why it was said above that nearly all functions of the AERB will be transferred to the NSRA.)

The Council will consist of the Prime Minister as its chairperson and will include six cabinet ministers [(including environment, external affairs, health, home affairs and Science and Technology (S&T)], Cabinet Secretary (ex-officio), the Chairman of the

AEC (ex-officio) and an unspecified number of eminent experts nominated by Central government.

The Standing Committee has said that, in order to make the functioning of the Council more effective, the number of Members of the Council needed to be restricted to a reasonable limit and not be left open as it is now. It has, therefore, recommended an upper limit of five on the number of experts to be nominated by the Central government. For this, it has suggested insertion of 'not exceeding five' in Clause 5(j) of the Bill.

The simplest of the functions of the Council is to constitute two search committees, one to choose the Chairman and the other to select members for the NSRA, and to fill vacancies as and when they arise (Section 10). More importantly, the idea of the Council was to bring to bear the perspectives of ministries such as health, environment and S&T as well as external experts, and have a comprehensive view across the spectrum of areas that nuclear activities impinge upon and evolve appropriate safety policies. The Council will also serve as a body to ratify various standards and codes that the NSRA will develop for different nuclear related activities. In the current dispensation, there is no external body that ratifies the codes evolved by the AERB. The Council will also be involved when disputes arise and Section 35 empowers the Council to constitute an Appellate Authority, to hear appeals against any order or decision passed by the NSRA, consisting of a Chairperson and not more than two members.

The Standing Committee has felt that the number of Members of the search committees, which will play a vital role, should not be left open to interpretation or discretion. It has, therefore, recommended that the composition of the search committees should be clearly specified in the Bill. It has also opined that an Appellate Authority—which will be a quasi-judicial body superior to the NSRA—consisting of a judge of Supreme Court or the Chief Justice of a High Court and two eminent scientists as Members would be an appropriate forum to hear the appeals of the aggrieved persons. The Parliamentary Committee also noted that Clause 35(4)(b) provides that a person shall not be qualified for appointment as a member of the Appellate

Authority unless he is an eminent scientist and has held the post of Secretary to the government of India or any equivalent post in the Central government but observed that there might be many more eminent scientists who would not have held the high posts mentioned above and the eligibility conditions as prescribed would deprive them of being included as members of the Appellate Authority. The Committee, therefore, recommended that the prescribed eligibility criteria should be broad-based and the Bill should be accordingly amended.

In principle, one could question the need for such a Council if the Authority was going to be vested with full autonomy. In fact, there have been criticisms already that a government appointed Council would render NSRA even weaker than the AERB and that having such a Council negates the idea of full autonomy to the NSRA. The argument being that a government Council appointed search committees to constitute the NSRA, could result in a pliant NSRA. But the point is, in the absence of a Council or an apex ministry, some other Government body will have to search and select. Also, an explanatory note to Section 8 makes the desired autonomy of the Authority explicit by saying, "For the removal of doubts, it is hereby clarified that notwithstanding anything contained in Section 7 [on the Council], the Authority shall be autonomous in the exercise of its powers and functions under this Act". But, more specifically, the Authority will not report its activities to the Council, but to the Parliament (Sections 38 and 39), including every rule and regulation that it promulgates (Section 52).

There have been criticisms against Section 48 of the Bill as well. It has been argued that it too is intended to reign in the Authority from exercising complete freedom and autonomy. Clause 48 (1) gives powers to Central Government to supersede the authority by notification for a period not exceeding six months, if in the government's opinion, the Authority has failed to discharge its duties properly and functions efficiently, and in accordance with the Act and Rules and Regulations made under it, and also when "circumstances exist which render it necessary in the public interest to do so." Though the provision requires the government to reconstitute the Authority within a notified period, Clause 48(3) allows for the extension of the period of supersession for further slabs of six months till a new Authority is reconstituted.

It has been argued that this provision gives the government arbitrary powers to dissolve the authority and replace it by a more pliant authority. The condition within quotes above is particularly likely to invite criticism. However, it should be pointed out , Section 48(1) does require that, before issuing the notice of supersession, "The Central Government shall give a reasonable opportunity to the Authority to show cause as to why it should not be superseded, and shall consider the explanations and objections, if any, of the Authority." Also, most importantly, Clause 48(4) requires the following: "The Central Government shall cause a notification issued under sub-section (1) and a full report of any action taken under this section and the circumstances leading to such action to be laid before each House of Parliament at the earliest."

In defence of the provision it may be argued that, in the absence of an apex ministry for the NSRA, this is required to guard against gross failures or inefficient functioning of the Authority in detriment to the safety of workers and the public. But, there are bound to be genuine apprehensions that such a provision is likely to be misused towards some ill-conceived strategic and political ends.

To ensure independence and autonomy of the NSRA, the Standing Committee made the following specific observation. Noting that the Bill met, by and large the three of the four core criteria required for the efficacy and success of the NSRA, namely competence, independence, stringency and transparency, it said that it lacked somewhat on the count of independence. The Committee observed that certain clauses in the Bill, namely Clause 14(1) on Removal of Chairperson and Members of the NSRA, Section 42 on Directions by the Central government to the NSRA and Section 48 on Power of Central government to supersede the NSRA, may impinge on functional autonomy of the NSRA. The Committee has, therefore, recommended that the DAE should strive to make Authority more independent and autonomous "not only to carryout its functions effectively but also to enjoy credibility among the public and the trust of the people".

However, though the Authority will not be under any ministry, its budgetary allocations will come from the annual budgetary grants of the DAE. For the year 2012-13, when the NSRA is expected to begin functioning, it is estimated that ₹ 37 crore would be required for establishing the authority, according to the Financial Memorandum appended to the Bill. For the entire 12th Plan period, the estimated budget is ₹ 164 crore 'for strengthening and expanding the activities of the Authority'. This may be compared with current AERB's annual budget of about ₹ 25 crore. This structure is somewhat similar to that of the Reserve Bank of India, which, though fully autonomous, comes under the finance ministry for its funding.

One could, of course, argue that the NSRA be made a constitutional body like the Election Commission or the Comptroller and Auditor General (CAG). Then the natural counter question would be, why not other regulatory bodies such as SEBI or TRAI? The point is that any autonomous body that is sought to be created has to be constituted by the government through an act of Parliament and the body will have to function within the guidelines stipulated by the government and be made answerable only to the Parliament and not to any ministry or another apex body.

So, what significant difference is the new structure likely to make and what is the way forward? Clearly, the formulated new regulatory structure is an independent body and not a subordinate of the DAE. So a more frank and transparent assessment of safety of the DAE/NPC installations is likely and the NSRA will not be constrained against taking suitable actions. Safety related data and environmental radioactivity release data would be more easily available in public domain. As required by the Bill, there will be regular audit of its functioning and detailed safety audit reports placed in the Parliament.

In any case, the recommendations of the Standing Committee have sought to address most of the criticisms against the Bill and these are in the process of getting incorporated at present.

However, the current ground reality of higher education in the country, particularly in nuclear science and technology, which is still confined to the DAE institutions only, the new body too will have to depend on the DAE for expertise to carry out its regulatory functions. For an effective and truly unbiased external safety regulatory authority, higher education and research in nuclear technology should become more widespread in the university system and R&D institutions. The Homi Bhabha National Institute (HBNI) of the DAE may succeed in realising that in the next 10 years or so.

So, it can be said that the proposed Bill is a first satisfactory step towards a truly independent and functionally autonomous to the nuclear regulator. This has to be an evolving process depending upon domestic and international experiences over time with increasing nuclear power generation in here and elsewhere (and associated safety-related events, if any) in the coming years.

## References

Gopalakrishnan, A. (1999). Frontline March 26.
Hindu (2011). "Review N-Plant Designs: Expert", *The Hindu*, March 20.
Ramachandran, R. (2000). *Frontline*, July 21.